THE FUTURE OF ORGANIZATIONS

Workplace Issues and Practices

21st Century Business Management

THE FUTURE OF ORGANIZATIONS

Workplace Issues and Practices

Edited by
Arvind K. Birdie, PhD

Apple Academic Press Inc.
3333 Mistwell Crescent
Oakville, ON L6L 0A2 Canada

Apple Academic Press Inc.
9 Spinnaker Way
Waretown, NJ 08758 USA

© 2019 by Apple Academic Press, Inc.

First issued in paperback 2021

Exclusive worldwide distribution by CRC Press, a member of Taylor & Francis Group
No claim to original U.S. Government works

ISBN 13: 978-1-77-463062-4 (pbk)
ISBN 13: 978-1-77-188623-9 (hbk)

Library and Archives Canada Cataloguing in Publication

The future of organizations : workplace issues and practices / edited by Arvind K. Birdie, PhD.
(21st century business management book series)
Includes bibliographical references and index.
Issued in print and electronic formats.
ISBN 978-1-77188-623-9 (hardcover).--ISBN 978-1-315-11489-7 (softcover)
1. Work environment. 2. Organizational effectiveness.
3. Service industries--Personnel management. I. Birdie, Arvind K., editor II. Series: 21st century business management book series

| HD7261.F88 2018 | 331.25'6 | C2018-902742-8 | C2018-902743-6 |

Library of Congress Cataloging-in-Publication Data

Names: Birdie, Arvind K., editor.
Title: The future of organizations : workplace issues and practices / editor, Arvind K. Birdie, PhD.
Description: 1st Edition. | Waretown, NJ : Apple Academic Press, [2018] |
Includes bibliographical references and index.
Identifiers: LCCN 2018021652 (print) | LCCN 2018023045 (ebook) | ISBN 9781315114897 (ebook) | ISBN 9781771886239 (hardcover : alk. paper)
Subjects: LCSH: Organizational change. | Work environment. | Job stress. | Sex discrimination in employment.
Classification: LCC HD58.8 (ebook) | LCC HD58.8 .F88 2018 (print) | DDC 658.3--dc23
LC record available at https://lccn.loc.gov/2018021652

Apple Academic Press also publishes its books in a variety of electronic formats. Some content that appears in print may not be available in electronic format. For information about Apple Academic Press products, visit our website at www.appleacademicpress.com and the CRC Press website at www.crcpress.com

CONTENTS

8.   Creativity and Innovation: The Future of Organizations....... 203

ABOUT THE 21ST CENTURY BUSINESS MANAGEMENT BOOK SERIES

Series Editor:
Arvind K. Birdie, PhD
Program Director PGDM, and Associate Professor,
Vedataya Institute, Gurgaon, India
Email: arvindgagan@gmail.com

CURRENT BOOKS IN THE SERIES

Employees and Employers in Service Organizations:
Emerging Challenges and Opportunities
Editor: Arvind K. Birdie, PhD

The Future of Organizations: Workplace Issues and Practices
Editor: Arvind K. Birdie, PhD

Other topics/volumes are planned on these topics:

1. Globalization and Emerging Leadership
2. Positive Psychology and Today's Organizations
3. Changing Consumer Behavior and Organizations
4. The Impact of Technological Advancement on Organizations
5. Emerging Employer and Employee Relations
6. Designing Future Organizations and Emerging Sectors
7. Aging in South Asia and the Impact on Emerging Businesses
8. Issues in Intercultural Management
9. The Role of Spirituality in Management
10. Increasing Workforce Diversity in Organizations
11. Creating Innovation in Organizations
12. Purchasing Power and Happiness in Customers

ABOUT THE EDITOR

Arvind K. Birdie, PhD, is Program Director of Postgraduate Studies and Associate Professor at Vedatya Institute, Gurgaon, India. Dr. Arvind has been consistently recognized for her teaching abilities. As an avid reader her strength lies in teaching various interdisciplinary subjects with equal ease. In addition to her academic teaching and training, she has organized various management development programs for corporates and academicians. She is a regular presenter at various international and national conferences and has published papers in refereed journals. Her areas of interest include leadership, work–life balance, virtual work and positive psychology. Dr. Birdie is the recipient of the Prof. Mrs. Manju Thakur Memorial Award 2016 for Innovative Contributions in Research/Test Construction/Book Publication for her book *Organizational Behavior and Virtual Work: Concepts and Analytical Approaches* (Apple Academic Press, 2016). The award was presented during the 52nd National and 21st International Conference of the Indian Academy of Applied Psychology, held at the Department of Psychology, University of Rajasthan, Jaipur, February 23rd–25th, 2017. She enjoys travelling and spending time with her family in free time.

LIST OF CONTRIBUTORS

Shaikh Abul Barkat
Maulan Azad National Urdu University, Mumbai Regional Centre, Navi Mumbai, India
E-mail: abulbarkat_2006@yahoo.co.in

Samo Bobek
Faculty of Economics and Business, University of Maribor, Razlagova, Maribor, Slovenia
E-mail: samo.bobek@um.si

Živana Veingerl Čič
Faculty of Economics and Business, University of Maribor, Maribor, Slovenia
E-mail: zivana.veingerl1@um.si

Madhu Jain
Associate Professor, Department of Psychology, University of Rajasthan, Jaipur
E-mail: madhujain28@gmail.com

Ruchi Joshi
Counseling Psychologist, Udaipur, India
E-mail: ruchijoshi43@yahoo.com

Purva Kansal
Associate Professor, University Business School, Panjab University, Chandigarh
E-mail: purvakansal@pu.ac.in

Navdeep Kaur Kular
Vedatya Institute, Gurgaon, India
E-mail: novikular@rediffmail.com

Subrata Kumar Nandi
Faculty Member, Marketing & Operations, ICFAI Business School, Pune, India
E-mail: sknandi2011@gmail.com

Lokesh Saxena
Managing Director, DISA India Ltd., Bangalore, India
E-mail: saxena_lokesh@hotmail.com

Manisha Saxena
Associate Dean (Academics) & Faculty Member, HRM & OB, ICFAI Business School, Pune, India
E-mail: manishasaxena75@hotmail.com

Nandini Srivastava
Faculty of Management, Manav Rachna International Institute of Research and Studies, Faridabad, India
E-mail: nandini.fms@mriu.edu.in

Girijesh Kumar Yadav
National Institute of Occupational Health, Indian Council of Medical Research, Ministry of Health & Family Welfare, Govt. of India
E-mail: girijeshkrydv@gmail.com

Simona Šarotar Žižek
Faculty of Economics and Business, University of Maribor, Razlagova, Maribor, Slovenia
E-mail: simona.sarotar-zizek@um.si

PREFACE

Organizations are changing at a fast pace with globalization and technological advancements. The world of work is in a state of flux, which is
causing considerable anxiety—and with good reason. There is a growing
polarization of labor-market opportunities between high- and low-skill
jobs, more women power by women entering in workforce, and more
technologically aware consumers. And from Mumbai to Manchester,
public debate rages about the future of work. Women represent one of
the largest pools of untapped labor/workforce globally; 655 million fewer
women are economically active than men. In a "best-in-region" scenario
in which all countries match the rate of improvement in gender gaps (in
labor force participation, hours worked, and sector mix of employment)
of the best-performing country in their region, $12 trillion more of annual
GDP would be realized in 2025, equivalent in size to the current GDP of
Japan, Germany, and the United Kingdom combined.

New concepts of learning organizations, flat structure, and knowledge
management for organizations are becoming in demand in today's business, and it is also necessary to be involved in innovation and developing
creativity in an ever-changing environment of a digital world without
boundaries.

Organizations in the past have metamorphosed into revolution all over
world. At the dawn of the 21st century, all highly industrialized countries
have become "service economies," at least when measured in terms of
share of the workforce employed in service industries.

More and more companies are service oriented, and with globalization
and technological advancement, expectations and skills required by organizations are changing at a fast pace.

The above-discussed concepts are becoming more prevalent with
advancement of technology and pressure to always evolve in these new
and changing times for business leaders and workforce. This book explores
the workplace issues and practices that will prevail, affect, and structure
future organizations, particularly in orientation of service organizations in
the 21st century.

This is the second book in the book series 21st Century Business Management, which aims to provide insights for implications and explores new basic concepts and principles of organizations and management studies in light of changing concepts as well as emerging concepts. In the first book in this series was on the upcoming challenges/opportunities for employees and employers in the themes of Generation Y, knowledge management, leadership effectiveness, work–life balance, spirituality, and emerging positive psychology concepts (such as psychological capital, knowledge management, mindfulness, particularly in the orientation of service organizations in the 21st century), were fully presented and explored.

The approach to developing chapters for books in the series is by reviewing and conducting empirical studies, round tables, and focus discussion with employees and employers of the service industry by the authors.

I would like to acknowledge with gratitude the support that I received from the contributing authors. Their rich insights and shared perspectives have enabled us to bring in many unique concepts of interdisciplinary ways to understand today's business world. We are indebted to organizations and many industry practitioners who provided support and discussed their ideas with the authors.

I would like to express my deepest gratitude to my parents, in-laws, my husband, Gagan, and our wonderful son, Shauryavir, for having supported me all along. I would also like to thank my brother, Permendra Singh, for his encouragement and always being there for me. Without their patience, support, and motivation this work would never have been completed. My family has always been a great anchor and always had a great role in all my accomplishments. Their being with me is the greatest joy, which enables me to make a meaningful contribution.

I deeply value the strong support I received from the extraordinary team at Apple Academic Press. A deep gratitude and my heartfelt thanks to Ashish Kumar, President, AAP, for believing in me to initiate the series and constantly encouraging me to accomplish this work. I deeply value his guidance. My special and heartfelt thanks to Sandra Sickels, Vice President, Editorial and Marketing, for support in terms of using her sensitivity and good eye to design the cover and flyers for the series. She was immensely helpful throughout the project. I also express special thanks to Rakesh Kumar for his valuable support.

To all these truly excellent people and to many others, too, I offer my personal warm regards.

CHAPTER 1

GEARING UP FOR THE FUTURE: HR TRANSFORMATIONS IN THE MANUFACTURING SECTOR

MANISHA SAXENA[1,*], SUBRATA KUMAR NANDI[2], and
LOKESH SAXENA[3]

[1]HRM & OB, ICFAI Business School, Pune, India,
*E-mail: manishasaxena75@hotmail.com

[2]Marketing & Operations, ICFAI Business School, Pune, India,
E-mail: sknandi2011@gmail.com

[3]Managing Director, DISA India Ltd., Bangalore, India,
E-mail: saxena_lokesh@hotmail.com

"The snake which cannot cast its skin has to die. As well as the minds which are prevented from changing their opinions; they cease to be mind".
—Friedrich Nietzsche

CONTENTS

ABSTRACT

The business landscape is changing rapidly and the competing environment is getting increasingly complex. The four major characteristics that define the existing business environment are volatility, uncertainty, complexity, and ambiguity (VUCA). These four dimensions pose significant chal- lenges for organizations, as they need to be fleet-footed in assessing the market situation and predicting the results of its actions.

An organization needs to formulate its strategies to adapt to these changes and constantly update them, as the rapidly changing business environment may not provide it with adequate time to decide when to change, and how to change. Thus, creating an agile organization that is capable of adapting itself to embrace this change requires well-thought- out strategy for its implementation. One of the determinants of an organi- zation's ability to navigate through the competitive environment is its HR policies that help manage their employees.

This chapter attempts to explore how some of the leading companies in the manufacturing industry are responding to the changes in the external business environment and adapting their HR practices in tune with these changes. It also captures qualitatively how these organizations foresee the future and the steps they are undertaking to prepare themselves for the future. The findings of the study are based on depth interviews with some of the HR leaders from the organizations in context.

1.1   INTRODUCTION

The human resource function within the organization has long remained under shadows of the other functions such as marketing and operations, perhaps because of the difficulty in quantifying its efforts. However, the changes in the external business and internal organizational environment are foreseeing effectiveness of human resource management (HRM) to likely impact the way a firm competes in today's environment.

Lepak et al. (2005) observed a shift from traditional bureaucratic organizational functions to more creative and flexible services by HR managers today to meet the demands of the changes in the internal and external environment. Such a transformation from the transaction and traditional domains to more transformational domains is the outcome of the expectations from the HR function to play a more strategic role within the organization (Beatty and Schneirer, 1997).

Agarwala (2003) observes that in order to effectively compete in today's business environment, in organizations the traditional set of standard practices needs to be replaced with new and improved HR policies. Changes in HRM in response to those in the business environment have already been noticed (Stroh and Caligiuri, 1998). Ulrich (1997) urges organizations to adopt innovative HR activities. Agarwala (2003) compares innovative HR practices to be similar to administrative innovation rather than technical innovation, as they are aimed at influencing attitudes and behaviors of employees.

With the emergence of new age competitors and rapid growth in technology, many organizations are facing a unique challenge—attracting and retaining the right people. It is generally accepted that people are key to the success of organizations. With the change in the nature of business and work, many organizations are looking at multitasking skills, emotional intelligence, and team competencies besides technical skills in employees. Coworkers and colleagues no longer share a common workspace, but may be distributed across the world, connected to each other through technology. Old concepts of management and leadership theories are no longer applicable and new concepts like lifelong learning, total quality management, authentic leadership, and ethics have emerged.

Organizations are undergoing transformations from organizational structures, to people skills to management styles, to type of leaders to work spaces, and to the work itself. In the midst of all this it is not possible for HR, which steers the whole organization, to remain and sail the boat in the same manner as was done earlier. The function of HR today is to not only think about today but also to look into the future to ensure that the organization is able to achieve its objectives of longevity and healthy growth. Managing people must be looked with a new lens. But are all firms and industries really responding to these changes by instituting appropriate changes in their HR policies and systems to have and maintain the right people within the organization?

The answer to the above question may not be easy, because each industry has its own peculiarities in terms of their nature of business and markets. The literature highlights the need for change in approach towards HRM in this rapidly changing business environment, and there is also evidence of companies who are adopting innovative HR practices. In this chapter, we attempt to look at the manufacturing industry in India and explore how some of the leading companies in this industry are gearing up for the future. Therefore, this chapter explores how some of the companies in the manufacturing industry have responded to the changes in the external business environment and adapted their HR practices in tune with these changes. The findings of this study are based on depth interviews conducted with HR managers from five manufacturing organizations in Pune, Maharashtra, to understand how the organizations in this sector foresee the future and the steps, they are undertaking while responding to them. This chapter is organized as follows. The first section sets the context of this research. The second section presents a review of literature highlighting the changes that have taken place in HRM and the linkages between HRM and organizational outcomes. The third section discusses the methodology adopted for the study. The fourth section discusses the findings of the study, and the last section presents conclusions.

1.2   REVIEW OF LITERATURE

Highlighting the challenging environment, which faces organizations today, Ulrich and Wiersema (1989) compared the times ahead as playing basketball with a moving basket. According to the authors, globalization and changes in technology, demographics and public policy have contributed to heightened organizational complexity. Thus, managing in these turbulent times require constant strategic adjustments and top managers ought to focus their attention on developing a "strategic capability"—the ability of an organization to think and act strategically—in a changing competitive environment and making adjustments to strategy. This requires a blending of analytical skills and administrative savvy; this capability can only be developed over time as there are no easy or quick solutions (Prahalad, 1983).

Even the small- and medium-sized enterprises (SMEs) are also found to be extremely vulnerable especially to the vagaries and turbulences of the external environment. Training is recognized as an important tool for developing their internal capabilities (Manimala and Kumar, 2012).

For the first time in history, the pace of change has produced four distinct generations (Baby boomers, Gen X, Y and Z) in the workplace. This has led to the need for caring manager, great coworkers, and a feeling of doing something, great leadership and a positive culture to avoid and resolve the inevitable conflicts as organizations look for motivated, enthusiastic, and productive workers (Branham and Hirschfeld, 2010).

One of the keys to successful development and deployment of HR service delivery applications is the recognition that these several generations at work interact with technology in very different ways. A "one-size-fits-all" approach to this challenge will limit one's success in HR service delivery efforts (Greene, December 2013/January 2014).

Beginning from simple personnel management to finally exploring the extent to which it is a source of competitive advantage for organizations, HRM has come a long way. The last two decades have been witnessed to these development, as a result, HR function has changed stance from reactive, prescriptive, administrative to proactive, descriptive, and executive (Boxall, 1996). Investigators have identified many important contingency variables influencing HR practices such as size of the organization, technology, age of the organization, presence of formal HRD, type of ownership, the existence of training units in HRD, and the life cycle stage of the organization (see, e.g., Jackson et al., 1989; Tayeb, 1988 in Budhwar and Khatri, 2001).

Best workplaces are categorized by continuous innovative ways to keep pace with technology leading to changes in the nature of work and diverse and dispersed organization's workforces (Levering and Erb, 2011).

Modern organizations are becoming increasingly diverse. The organizational diversity is common, and it is acknowledged to have competitive implications. Martin-Alcazar et al. (2012) conceptualize diversity as a multidimensional construct comprising demographic diversity, value, experience, and cognitive and knowledge diversity. The elements of demographic diversity include age, gender, nationality, education level, training, tenure, and functional diversity. With workforce diversity in organizations increasing, firms need to manage such diversity. Stockdale and Crosby (2004) define diversity management as a planned voluntary program, which exploits diversity among employees to become sources of complementarity, creativity and heightened effectiveness. Managing a diverse workforce requires a complete reconsideration of strategic HRM (Syed and Kramar, 2009). Martin-Alcazar et al. (2012) in their study found that certain patterns of strategic HRM practices promote collectivism, inclusiveness and appreciation of individual differences, and influence the effect of diversity on performance.

In recent times, there has been an increasing interest on work–life balance (Pocock et al., 2005). It is assumed that for organizations to truly engage with their employees and keep them motivated, they should be sensitive about the work–life balance of their employees. Several factors which have contributed to the interest in the issue of work-life balance include technological changes, global competition, aging workforce, demographic changes, and changes in values (Gunavathy, 2011). Work-life balance is no longer restricted to women workforce, and is today relevant for all types of employees (Bird, 2006). According to Felstead et al. (2002), four theoretical positions—institutional theory, situational theory, high commitment theory, and organizational adaptation theory—explain the factors which guide organizational practices related to work-life balance. Institutional theory assumes that organizations undertake actions to display social legitimacy, which are governed by normative pressures in the society. Situational theory suggests that organisations' adoption of work-life provisions are pragmatic responses to talent acquisition and retention challenges. High commitment theory assumes family-friendly employment practices are linked to human resource strategy of a firm, which aims to improve employees' organizational commitment. Finally, organizational adaptation theory suggests that organizations in addition to normative social pressures, consider workforce characteristics, work organization, and management attitudes while explaining the work-life balance initiatives by a firm. Work-life initiatives can take the form of money-based, information-based, and time-based strategies (Thompson, 2002). Money based strategies include childcare subsidies, childcare tuition discounts, and leave with pay. Information-based strategies include dependent care resources, relocation assistance, and referral service. Time-based strategies include policies like job-sharing, telecommuting, compulsory vacations, compressed work weeks, and flextime. All these factors contribute to the employee motivation and help enhance employee commitment to organizations. Yet, the response of organisations to work–life balance is not uniform, as reported by Gunavathy (2011).

Such changes in the environment forced organizations for optimum utilization of resources for their performance with the help of capacity building through HR function. One such study in an automobile component manufacturing organization used a developed and validated instrument to derive excellence in three HR functions: R and S; T and D, and PMS (Krishnaveni and Sripirabaa, 2008).

According to Boselie et al. (2005), HRM is concerned with the choice of practices, policies, and structures that organizations use to manage their employees. All the decisions and activities of the management impact the relationship between the employees and organization. Therefore, HRM encompasses a variety of practices and policies like design of tasks; staffing including recruitment, selection, career development, and so forth; performance and reward management; and facilitating channels of communication for work participation and decision making.

The changing business environment forces a relook into the field of HRM, its impact on HRM and the contemporary developments in this area to make the right business impact. This involves leveraging HR technology from induction to exit through innovative HR practices that attract, retain, develop, grow, engage, align and help in recognizing aspirations, and transitions of employees.

Research indicates that organizational performance can be enhanced by appropriately designed HR practices (Nishii et al., 2008), which influence the attitudes and behaviors of employees (Wright et al., 1995). Since the 1980s, there is a growing realization among organizations about the linkages between HRM and business performance. De Leede and Looise (2005) propose a model where it is shown that organization strategy guides HR practices, and HRM practices influence HRM outcomes like motivation, commitment, cooperation, turnover, and so forth, which in turn impact organizational outcomes like profits, productivity, quality, and so on. Some of the HRM outcomes have become very critical for competitive survival of firms. For example, lack of employee motivation and commitment and high employee turnover can have a detrimental effect on a firm. Therefore, HRM should adopt appropriate practices to achieve desirable outcomes.

Hence, it is imperative for companies to keep their firm strategy in mind while designing their HR practices. Firms generally organize human resource practices into systems that are consistent with their culture and business strategy (Osterman, 1987).

Best Practices (Universalistic Approach) adopted across different industries (Tomar, 2011):

1. Employment security
2. Recruitment and selection practices
3. Employee ownership of tasks and work modules
4. Sharing of organization level information with all employees
5. Employee participation in organization level policy-related decisions
6. Employee empowerment
7. Frequency of job redesign
8. Opportunities for training and skill development
9. Emphasis on cross training of employees
10. Presence of cross-functional teams
11. Symbolic egalitarianism
12. Wage compression
13. Promotion systems
14. Adoption of 360 degree appraisal process
15. Feedback mechanism

Many companies that seek competitive success achieve it through people that involve fundamentally altering how we think about the workforce and the employment relationship. It means achieving success by working with people, not by replacing them or limiting the scope of their activities. It entails seeing the workforce as a source of strategic advantage, not just as a cost to be minimized or avoided. This different perspective helps firms to successfully outmaneuver and outperform their rivals (Pfeffer, 1995).

Literature evaluating the link between organizational strategy and HRM can be classified into three different schools of thoughts: the contingency approach, the best practices approach, and resource-based approach. Delery and Doty (1996) term these three perspectives as universalistic, contingency, and configurational. The contingency or "best fit" approach considers models, which advocate the need for aligning HR strategy according to the context in which the organization exists. The best practice approach suggests that there are certain best practices, empirically proven, which firms can adopt to improve its performance (de Leede and Looise, 2005). The third approach, the resource-based approach, considers valuable resources as those which are inimitable and non-substitutable. These resources can enhance organization's capabilities and competencies. Thus, human resources of a firm have the potential to become valuable resources helping firms gain competitive capabilities.

In recent time, there has been a lot of interest among researchers on perceived organizational support (POS). POS is the degree to which employees believe that their contributions are valued, and the organization is concerned with their well-being (Rhoades and Eisenberger, 2002). Organizational support has assumed importance in the wake of the rapid shifts in the world of work (Baran et al., 2012), triggered by rapid globalization and advances in technology (Burke and Ng, 2006). Employees' POS is influenced by his socioemotional needs and how readily the organization rewards him for increased efforts that the employee makes on behalf of the organization (Rhoades and Eisenberger, 2002). Studies indicate that POS is directly related to various aspects of employee well-being and health. High POS has been found to improve general health, reduce burnout and anger, and induce a sense of contribution and accomplishment (Baran, 2012).

Thus, we can summarize that HR is trying to respond to the changes being witnessed in the environment. To substantiate the outcome of the literature review, the study was designed to study the manufacturing sector, which seems to be left behind in area of HR practices. Hence, objectives were set for the same.

1.3 OBJECTIVES

The nature of workforce in an organization often determines the capabilities of an organization to compete in an increasingly turbulent economic environment. It is generally acknowledged that effective management of human resources is a key to organizational success. Mesch (2010) observes that human element is the most fundamental among all the factors, which contribute to organizational performance. Every organization desires to have a set of highly motivated and committed workforce. To achieve this desirable end, organizations need to have effective HRM.

This chapter brings to the forefront the undergoing unprecedented changes in the business environment in India and the need of the human resource function of organizations to keep pace with the transformations taking place in the external environment. The manufacturing sector seems to respond slowly in the context of these changes, especially in the last two decades. The study intends to address this gap. Thus, the objectives of the study are as follows:

1. Understanding the HR transformations taking place in the manu-facturing sector in India
2. The challenges faced by organizations today in managing their HR
3. The strategies that organizations are adopting to manage their HR
4. Key policies adopted by these organizations in the areas of recruit-ment, selection, retention of employees, performance measure-ment systems, employee engagement, and so forth.

The questions in the interview schedule (annexure) were framed based on the above objectives.

1.4 METHODOLOGY

The study was exploratory in nature and intended to document and describe the phenomenon of interest (Marshall and Rossman, 2006). To achieve the aforementioned objectives, a schedule was prepared for an interview with the HR heads of organizations belonging to some of the leading manufac-turing firms, post which the data was analyzed qualitatively. The interview questions included assessment of the challenges faced by the HR func-tion by the industry, in general, and the organization in particular, and the responses of the organization's HR function, and so forth.

The interview responses have been supplemented by information available in the public domain through documents, such as, the company annual reports, newspaper, magazine and journal articles, and also infor-mation available on the company websites. This was done after the review of related literature helped in building some insight. The research elicits tacit knowledge, subjective understandings, and interpretations and delves in-depth into innovative HR processes trying to search the informal and unstructured linkages as relevant variables have yet to be identified. It is completely justified that a qualitative research is conducted. Moreover, the intent is to understand what different HR processes are being adopted in the midst of aggressive changes in the environment. The changes are assumed to address various behavioral concerns of the employees in terms of their satisfaction, performance, retention, motivation, and so forth. This does not call for coding the social world. The particular setting for the study is the workplace in manufacturing sector, and the sample is set of people from the HRD.

The sample had cases that met an important criterion of being the leaders of the manufacturing sector who would set benchmarks for others to follow. It was also a convenience sampling along with the same which helped save time and effort. The authors adopted purposive sampling with regard to the selection of organization. The respondents comprised the middle level and senior level employees in the organizations chosen.

1.5 FINDINGS

This section presents the findings of the study. As explained in a previous section, the data was collected through structured interviews with HR heads of five leading manufacturing organizations representing auto component, energy, and power equipment industries. Needless to say, all the interviewees are of the view that the HR function has become more challenging due to changes in the external business environment. Two major challenges for these companies are attracting and retaining talent. In an increasingly globalized business environment, the workforce is faced with increased job opportunities. Employee loyalty towards the organization has rapidly reduced over the last few decades. In order to overcome these challenges, these firms have adopted appropriate changes in their HR practices and policies.

This section discusses the changes encountered in the organizational and business environment by firms and the best HR practices adopted by the organizations which have been part of this study.

All the participants in the interview were of the opinion that there is a severe talent shortage both at the technical level as well as the leadership level, despite the availability of a pool of people who have the right qualifications. Quality manpower is difficult to identify and recruit, as few from this pool have the right skills and attitudes to perform the tasks assigned to them. As a result, the role of HR becomes critical, as it is not only important to have people who fit in the organization and the job role, but also to keep the manpower costs within the budgeted figures. Existing leaders are finding it difficult to cope and hence organizations have moved their focus from looking for an intelligent leader to one who can think in the moment and be innovative. "We can't change the environment so we have started working on changing the leadership style" says a senior talent manage-ment head of one of the companies. This initiative is to have personality type that can deal with this ambiguity and uncertainty as organizations by and large are grappling with the situation to avoid stagnation. Further she comments, "What got you here will not get you where you want to be in the future."

Every decade witnesses some turnaround changes that need to be addressed by organizations to function smoothly. The future organizations are slated for huge changes including the manufacturing sector, thus they need to reinvent the wheel immaterial of the sector they belong to. The management of these organizations believes that technology will be the biggest enabler as well as a challenge for HR as lakhs of jobs would be lost to automation; use of drones, digitization, and so forth but the ultimate focus should still be on the R-3 (Recognize, Reward, and Reprimand). Trust is the name of the game. Collecting bargaining is a passé.

Because of the shortage of appropriate talent and availability of multiple employment opportunity available to people due to the presence of large number of firms, managers are experiencing changes in mone-tary and nonmonetary expectations of their employees. At the same time, they are also witnessing the change in their priorities and demands. From asking for a favorable culture to no compromise on safety both physical and psychological, the employees are demanding both hygiene and moti-vating factors. The kind of people joining these organizations today are the ones who are above their basic necessities and hence bother more for higher order needs as depicted in the Maslow's hierarchy of needs. They are keen to understand the direct link between their performance and the growth of the organization. With changes in the workforce, more women, youngsters, and so forth, the needs have changed. Youngsters are less worried about tomorrow, value instant gratification, work–life balance and personal space. With all this, paternalistic style of leading has outlived its utility. The average age of employees in most organizations has drasti-cally reduced over the years. For the sample organizations under study, the average age of employees ranged from 28 years to 37 years. Therefore, it has become extremely difficult to deal with such employees in terms of understanding their expectations and factors that motivate them.

Organizations do not have a right answer per se but need leaders who can lead and deal with existing changes. To incorporate the same, a lot of concentration is on the competency development and the methodology adopted for the same is mere experiential. While HR systems are in place, there is a drastic shift from a command and control system to a more supporting and counseling system where the attempts are being made to offer a structured direction to supporting employees and giving them opportunities to be innovative, and offer autonomy, freedom, and empowerment.

Types of engagements have changed as motivations have changed and hence the tools too should change. The industry is offering flexibility to keep pace with the need of the existing set of employees, and has facilities like work from home and flexible timing. But many Indian manufacturing companies are still coping with it. Most of the best practices come from the software companies as they have more opportunities to practice and implement the same, for example, Google; they also have a consulting approach as people would just like to be connected with the organization as an outsourced work rather than being engaged with them on a full time basis.

The organizations need to pay attention to what works for the younger generation. As workforce becomes more diverse in terms of generations (X, Y, and Z) gender, age, and ethnicity, things like crèche, good support in organizations, freedom, leaves, sabbaticals, or education leaves are also becoming common in Indian corporate organizations. Organizations aiming to gain competitive advantage are looking for innovations which demand a lot of support and freedom for creativity.

One of the participants was of the opinion that structural changes taking care of external environment leads to internal upheaval and chaos further giving rise to coping issues among employees who are still part of the same organization. Finding leaders to lead that change and right people to do the new allocated job are challenges because existing people are not equipped, and the skill sets are not readily available in the market.

HR manager in the PSU (Public Sector Undertaking) organization feels that generation Y just are more inclined to quick fix solution, and they seek to spend money on projects and look for speedy progress. A new emerging challenge is their perceived threat from automation. "Manufacturing is the heart of any economy, it is the key to growth in any economy, IT or technology can only be a facilitator" according to the PSU HR. It was reiterated that too much of technology penetration has hampered the one-on-one communication ability. Research orientation in employees is also lacking. Some of them were of the opinion that data-driven jobs might just make some managerial skills redundant. Leisure management is becoming a new area of research as people do not know what to do with their leisure time.

Though the ability to understand the system of management practices is hindered by its very extensiveness, whatever change needs to be brought has to be more comprehensive. The following 13 policies and practices emerge from the review of popular and academic literature, and talking to people in firms in a variety of industries. It is important to recognize that the practices are interrelated: employment security; selectivity in recruiting people who could work best in the new environment, could learn and develop, and need less supervision; high wages; incentive pay, employees, who are more inclined to take a long-term view of the organization; information sharing; participation and empowerment; self-managed teams; training and skill development; cross-utilization and cross-training; symbolic egalitarianism; wage compression; promotion from within.

HR heads of a few of the best companies (PSU, MNCs, and local family owned big group) in the area of manufacturing in Pune were contacted and their best practices are discussed here. The responses to various questions have been compiled and presented under various HR policies/practices/ activities (sequentially) in the next section.

1.6   BEST HR PRACTICES ADOPTED

1.6.1   RECRUITMENT AND SELECTION

Recruitment and selection have been defined as a positive and negative activity, respectively, as one is a process of inviting applications and the other of eliminating. It forms the basis of HR capital in any company and thus a serious thought for the same is most imperative. Just like an organization decides for the purchase of capital so does it brainstorms for selection of human capital.

Moreover, the organizations are trying to keep pace with changes and constantly devising novel ways of bringing the right person on board. They are reaching out to the prospective employees in more than one ways.

All the companies covered in the study hire both fresh graduates and lateral hires. Internal job postings ensure each and every employee has equal opportunity to apply without compromising on the merit. The organization makes a conscious effort to train and develop these people. In most of the cases, hiring at managerial and above levels are done through internal promotions. This provides growth opportunities for deserving candidates within and also ensures that a large percentage of key positions are filled by people who understand the culture and values.

In cases, where internal candidates aren't available and/or are not ready, the company looks at external hires with the help of recruitment consultants whose performance is evaluated on predecided parameters. Here, the objectives are twofold—one is to bring in at least a few people from outside who can challenge the current practices and processes, and innovate on the same. The other is to get in people with similar values so as to be consistent in their approach of how to get things done. Some of the evaluation parameters include: flexibility, innovation and creativity, adaptability, and attitude apart from basic competencies of knowledge and skills. Personality job fit and culture fit has become the new mantra.

The number of relevant profiles received for every position, time taken for the search, ratio of the profiles received to candidates selected, help with preemployment formalities, and so forth, are some of the deciding factors. All vacant positions allow employees the option of referring suitable candidates. There is no special concession given to internal referrals.

Concepts like hire to develop apply for a fresh graduate which implies that the relationship is a long-term one. From here only, the companies look for good regional and cultural blend and diversity of candidates. The focus here is to ensure that only fresh talent is hired who are clearly seen as future leaders. Their hiring process (recruiting, interviewing procedures, and so forth) ensures that a job candidate will fit into their culture. It believes that it is building its future on proactive people who are team players, with an open, innovative mindset; looking for a career in an organization that encourages empowerment. Their hiring process plays an important part in ensuring that they have the right kind of people, with relevant skills, knowledge, attitude and values, and ethics, in the appropriate roles so as to enable them to get the best out of them. One of the companies mentioned the importance of this by stating that the current MD is the product of 15 years of retention.

Keeping in mind the young, Gen-Y (people born during the 1980s and early 1990s) prospective employees, innovative means of communication are used. Extensive use of social media avenues such as official Facebook page have ensured that the company can reach out to and keep engaging them. In one of the companies interviewed the recruitment of each batch of trainee starts a year in advance, and various mediums have been employed to engage them to create a sense of belonging before they join—such as: sharing successes, newsletters, inspiring videos, one-on-one connect through social media to network, and many more such avenues. The company connects with these students all year round, and not only during placement season through other initiatives. The summer internship program is another process where talent is picked up much before campus hiring.

1.6.2 ORIENTATION PROGRAM AND PLACEMENT

Companies welcome new employees and make an effort to integrate them into their culture as it realizes the importance of a good start thus emphasizing on participation and enthusiasm among new recruits during their orientation program. They ensure that all new employees feel comfortable when they join, settle down quickly and feel part of the family.

The onboarding process is a crucial part of new hire to the organization which helps to have fully engaged and highly productive employees by integrating them into their new work environment and assimilating them into the organization's culture and advancing them to the desired level of productivity as quickly as possible. It is critically important that they are supported with the foundation tools, resources, and organizational perspective that ensure their rapid inclusion within workforce and culture. This onboarding procedure is run by all the HR business partners (HRBP) for their respective divisions.

Similar to the two-phase hiring process, induction programs are customized for lateral hires and fresh graduates separately. The guiding theme of one of the program is to provide the new employees with the "Company Experience". The participants get to interact with leaders, customers, and business unit heads, experience the CSR activities among other things. The objective behind this program is to have an institutionalized mechanism to foster one culture, promote commitment and sense of belonging, provide a platform for cross-pollination and networking of employees from various areas, and to acquaint the new employees with the big picture (strategy, business, customer, competition, and so forth).

It is ensured that these graduate engineer trainees and management trainees are given a wholesome experience by giving them exposure throughout the first year.

Experiential learning (visits to suppliers, customers, and sales field visit); interaction with leaders; role play-based learning, movies, world cafe, outbound events, and team building activities; on-the-job training; e-learnings, knowledge sharing sessions, action learning sessions are some of the ways how companies have adapted to the learning styles of Gen Y. This is done to gauge the new employees' interest along with his competencies, so that he gets an ideal job role at the end of the year.

The talent management team makes a point to stay connected with all the new employees through social media. This is done right from the day of selection, with an objective to keep the new employees updated with the company news and help them connect with their fellow new employees in order to encourage a strong bond between them.

In addition to that, one of the companies groups the new hires for a period of two years to help them settle down. These groups are of different functional areas and known as affinity groups. They also have similar group by the name of women affinity groups (WAG). A steering committee is there to manage the same.

1.6.3   LEARNING AND DEVELOPMENT

In the kind of environment that the organizations find themselves today, only a learning organization can get best synergies across the company. Current needs are mapped and developmental needs are identified in the due course of the year. Customer requests are also looked into by the HR partners and talent management team. In order to bridge the gap, customized in-house programs are designed and conducted.

Creating learning that sticks is mainly about creating a learning process that stretches beyond the classroom experience and into the daily work. The focus of the learning experience is around the learner's true business issues.

The objective is to elevate and develop the critical and strategic competencies of the designated job families, competency areas, and work and to create a global cross knowledge work group, to identify and develop critical skills. One of the companies has its own College Campus in varied global locations that reduces the need for traveling long distances away from work and family, as well as the high cost of hiring foreign trainers.

With this facility, one can study at their own pace, in their own time. All courses are web based so all that is required is access to a computer, a macro-media plug-in and a connection to the Internet. All courses are free of charge. One can choose to do all or only part of a course at one time. Courses can also be done several times. E-learning means "learning and teaching with the support of different electronic media". This means the traditional "Self-paced E-learning," but also other delivery methodologies, such as videoconferencing, audio conferencing, podcasts, apps, and so on.

Based on the need of the employees/organization, various specialized programs are designed and deployed using analyze, design, development, implementation, and evaluation (ADDIE) model.

Learning implementation plan (pre; during and post joining) is developed in order to create the individual Learning Implementation Plan. Training needs are generated from the annual company kick off and the BU/functional kick off. Country management team (CMT) shares the score card, and the leadership team discusses the goals and objectives for the current year. The development plan for the organization is a part of the score card and is discussed during this meeting. A broad agreement is reached on the areas to be focused for competence building and people development. Specific attention is given to identify leadership competencies that are critical for the future. The leadership engine is to focus on developing forward thinking in leaders and create a talent pipeline.

Companies have strategic partnership for higher education programs as they think about leadership for today and tomorrow. Leadership for growth (LFG) is in line with HR vision "think business think people" which is committed to enable companies to deliver for today and tomorrow, to accelerate the development of consistently high performing individuals with a potential to be "leaders" for business critical roles. This will ensure readiness of leadership capability for the future by building a strong succession pipeline.

Utilizing different technological tools and methods like web conferencing, e-learning, classroom setting, group work, project work, and coaching—the learning and development programs aim at enhancing the quality in learning for employees. Manager's involvement and support through coaching is important for employees to achieve a sustained personal, professional and behavioral development.

The employee performs a competency assessment to evaluate their own competencies (behaviors, skills, and knowledge). The manager and eventually peers and reports do the same. This is an input to the review meeting when manager and employee discusses and agrees on the development objectives and activities for the individual development plan (IDP) which is continuously followed up, especially in the face-to-face meeting in the end of the year.

1.6.4 TALENT MANAGEMENT

All the above might go in vain if the companies are not able to manage the talent and they have to bring in. Nowadays, companies believe in talent management rather than talent buy. As part of talent management process employees are rotated into different roles based on individual interests and organizational needs through a job rotation process. Job rotation aids in gathering the nuances of different jobs in the organization. This is one of the tools and policies to leverage global job opportunities. Positions are advertised and employees are encouraged to apply for open positions and selected through rigorous selection methods.

These companies give equal opportunity in a global environment and promote to develop global leaders. The implementation of IDP is an entitlement for all their employees and a catalyst to self-empowerment. Performance reviews, work climate analysis are further enablers.

Hire to develop policy of one of the companies has had an impact to the extent that their current MD is a product of 15 years of experience with the company. Initial settling down time for freshers is anticipated as 5 years. The company is known for fair treatment and has national and international tie-ups for MBA and M. Tech courses. It not only encourages but also sponsors employees for higher studies. This helps both in career planning and succession planning. It has gone to the extent of erasing the concept of fixating people based on their qualifications. For example, a CA will not necessarily get into a finance role if s/he feels HR is his/her calling. Employees who have shown inclination have been shifted and experimented with new roles which are entirely different and also have proved themselves successful in different leadership roles.

This organization has further stepped into the development and grooming of their blue-collar workers where decision-making is taught, education given, and communication and English speaking is taken. All this brings a difference in their lives.

"This is the ability of our leaders to take risk on people and give right direction" according to the HR head of the company.

1.6.5 EMPLOYEE ENGAGEMENT

Employee engagement is the buzz word today which has also become one of the key retention strategies. Such employee-friendly HR policies help to reduce absenteeism and employee turnover. Telecommuting (work from home to provide for work–life balance), flextime, birthday gifts, marriage gifts, farewell gifts, Diwali voucher, and so forth are some of them. Further a caring environment adds on to it. Every location is free to celebrate festivals and occasions.

The company not only celebrates team and/or organizational successes but also encourages fun and camaraderie among employees. Celebrations come in various forms, be it an event for awards, customer recognition, project completion, or any other festivities.

Local toastmasters club was started to provide a positive learning experience through which members are empowered to develop communication and leadership skills, resulting in greater self-confidence and personal growth. Toastmasters International is a world leader in communication and leadership development.

Table tennis, volleyball, cricket, and throwball tournaments instill team spirit, camaraderie, and at the same time drive a competitive mindset. Participation as a company in key events which are some form of charity, fun and at the same time develop a spirit of camaraderie for example, participations in Marathons, Stepathlon, and so forth

The employee engagement activities are decentralized and much more than mere fun events. A company mentions that their list of holidays honors all religions and thus is a testimony to its effective diversity management. All festivals are celebrated and respected: Diwali, Eid, Christmas, and so forth.

Another unique concept is identifying groups for leadership interaction, for sponsoring education, recognition, bringing families to work place, celebrating birthdays, and personal milestones. A regular thank you goes a long way in saying that you care. This is over and above the remuneration and the performance linked pay.

When evaluating performance—both short- and long-term—what is considered is also the results that have been achieved in line with the code of conduct as well as the company vision, mission, drivers and shared values and that the employee has an aptitude for teamwork and a "One" mindset in place. When assessing performance for managers, both achievement of business result as well as development of the organization and their reporting employees are weighed in.

The company shows appreciation and/or recognition for employees' good work and extra effort, or other achievements through awards.

1.6.6 CORPORATE SOCIAL RESPONSIBILITY

The organizations are very conscious of their role in conserving the environment as manufacturing has a direct impact on the environment. One of the companies has global ISO 14001 environmental certification and global health and safety management standard OHSAS 18001 certification. Its operations are also certified in accordance with either ISO 9001 or applicable customer segment standards, for example ISO/TS 16949 (Automotive), AS9100 Aviation or IRIS (Railway).

The focus is to reduce the environmental impact of an asset during its lifecycle, both in their own operations and their customers' operations with sustainable development in mind through not only business care but also environmental, employee and community care.

Business ethics is an important element of business care. Offering better products and solutions, which in turn help to improve their operational or product performances; committing to fight corruption and promoting ethical business behavior in all countries where the company is active. "The fight against corruption" is a group-wide awareness training originally produced by the United Nation. The manufacturers have gone to the extent of involving their suppliers to adhere to and adopt the code of conduct, environment, health and safety and zero defects concepts.

Thus, they not only restrict to supporting environmental friendly procedures in their own operations but also drive down this point to its suppliers, that is, maintain an environmental discipline across its supply chain.

The Leadership in Energy and Environmental Design (LEED) standard—or equivalent—is mandatory to apply for new major construction projects (over 1000 m^2).

One of the companies initiated a sports academy for underprivileged in July 2005. The objective of the Sports Academy is to provide children in Pune the opportunities to both physical and mental growth and development through sports activities and mentorship program.

Based on the four core values: high ethics, empowerment, openness, and teamwork; the company is committed to offering safe and nondiscriminatory working conditions to its employees. It is their obligation and responsibility to treat all employees fairly, equally, and with respect regardless of their race, gender, age, national origin, disability, caste, religion, sexual orientation, union membership, or political affiliations. Companies constantly encourage their employees to volunteer their services for social work in whichever way they deem fit.

1.6.7   WOMEN IN WORK ENVIRONMENT

Women make up about 47% of the labor force, but only 27% of the manufacturing workforce[1] in US. Out of 24% women as a working population, only 12% are in the Indian manufacturing according to the companies. Creating an enabling environment for women to succeed at workplace is the agenda for action in these companies, actively driving the principles of "Mix with Care" and "Care for the Mix" and has been to exploring avenues for recruitment, addressing special needs of women and creating levers for them to excel. Hiring freshly graduated women is a priority. In the last few years, more than 10% of the total fresh graduates hired, have been women. Another priority is to broaden the scope of sourcing women candidates in areas which have higher women representation such as legal, purchase, finance, and customer service.

Keeping in mind the developmental needs of women at different stages of their careers, one of the companies has drawn a three step developmental process (Motivate-aspire for bigger roles; Strengthen-threshold level by building existing capabilities; Anticipate-larger contribution for future) for enabling women to contribute effectively at the workplace and paving the way towards future readiness through developmental interventions.

[1] U.S. Bureau of Labor Statistics, 2014.

This aims to help women discover their strengths, become aware of the potential barriers, and help to develop a strong personal brand. Some of the enabling policies are maternity policy, day care center, sexual harassment committee, and so forth. There is women representation in almost all the functions. Chrysalis—the women's development program in one of the companies is about strengthening self-belief and enabling women to navigate effectively the personal and professional maze.

The objective is to empower women and create a gender balance. Moreover, it emphasizes diversity which brings creativity and culture of innovation. Apart from women empowerment initiatives and bringing in workforce diversity one of the organizations has gone for full inclusions that includes respecting people's preferences when it comes to lesbian, gay, bisexual, and transgender (LGBT).

The belief that an organization should be a reflection of the society has led to focused hiring of women and growing women leadership. Thus, the gender mix is witnessed at all levels from entry to higher.

1.6.8   SOME MORE INTERESTING HR POLICIES

Some of the companies have modified HR policies and adopted initiatives to support employees at times of significant life events—a personal crisis, family illness, birth, marriage; health checkups, hospitalization cover for the employee and the immediate dependents covered under a cashless scheme of health insurance policy arranged through a leading insurance company; medical assistance and support during death while in service policy—for employees and their families through Employees Contribution Towards Death Fund and other aspects (provident fund, gratuity and superannuation balance, encashment of privilege leave in the employee's account, and so forth); Employees Group Mediclaim Insurance Policy; parents mediclaim policy; group term life and group personal accidental death insurance. All this is apart from career enhancement initiatives, work–life balance through various types of restructured leaves (including paternity leave/adoption leave/compensatory off); flextimings; work from home/telecommuting; part-time assignments; canteen (breakfast, lunch, and evening snacks available at nominal charges); official mobile phones, laptops with latest and updated versions, and so forth. Day care center;

gymnasium, tennis court, throwball/volleyball court, swimming pool; and so forth are some of the facilities that also help in work–life balance.

Organizations realize that their responsibility towards employees is to respect them and their rights; to offer safe, nondiscriminatory, good working conditions; and to continuously develop skills and competencies to enable the individual's satisfaction and career possibilities.

The ultimate purpose of all these interventions, which goes unsaid, is attracting, motivating and retaining the best in class. The intent of this research was to give insights into the newer version of HRM.

Some of these facilities are hygiene factors (maintenance factors) from the Herzberg's two-factor theory of motivation. All this helps in reducing absenteeism, employee turnover. The past record for turnover has been minimum. Also short-term and long-term incentives along with various employee engagement activities keep the employees motivated. At the best workplaces, leaders also share the personal side of information such as emotional reactions to news, personal takes on values, or simply their hobbies and interests.

One of the organizations has recently stopped the punching system to check the log in hours as they believe trusting employees is the way forward. They have a performance-driven culture which is result-driven rather than emphasizing on best effort. So no punching, only concentrate on your deliverables 5 days a week. They are more concerned with the final productivity and outcome rather than just log in hours. They have come up with an innovative concept of collaborative work places for the past 2 years (2014 onwards) which means no fixed cubicles. The physical infrastructure has no defined and designated places, anyone can come and sit anywhere and work. This includes people from the senior management as well who demonstrate in their demeanor and walk the talk by walking out of their cubicles, interacting with all and changing their places. Such a system offers open channels of communication, gives space to people to voice their apprehension. Collaborative programs are available for employees to make suggestions. One of their values is about openness and transparency. This translates itself into open offices, open door policies, addressing employees by their first name and other elements which ensure everyone's voice is heard. As one company HR head pointed that we air on radio live chat of our employees in the evenings in their campus.

One of the companies encourages its employees to address each other by names, irrespective of their organizational hierarchy.

1.6.9   OUTCOME: MOTIVATED AND PRODUCTIVE EMPLOYEES

Though no direct empirical study was done to establish a relation between the innovative HR practices and productivity of the employees, but some criteria can be studied for the same. Low turnover can be one of the indicators. Over the last 5 years, more than 70% of the positions at Manager and above level have been filled by internal hires. Equal opportunities, transparency, openness, freedom, and recognition are the key characteristics of these companies. All these provide greater meaning to employees as they see their roles as extensions of their own values and needs.

Degree of responsibility of assignments handled by employees even at entry levels is much higher due to the flat reporting structures. Also, before finalizing on the company policies, all employees are asked for their suggestions/ideas on the existing policies. Most of these suggestions are incorporated while finalizing the policies.

They provide tremendous learning opportunities to employees throughout their working career. Management, especially senior management, shares information with employees and fosters a culture of transparency through distinctive ways and vice versa, that is, employees can ask questions, provide feedback, or otherwise communicate with managers, especially senior managers in different ways through Employee Forums, Working Climate Analysis. Intranet communication is accessible to employees, which enables employees to review success stories, locate experts and find solutions.

Some actions have also been undertaken to achieve zero accidents at work, where all units must have health and safety committees established with management and employee representation. All employees must go through health and safety training, hazard and risk assessments of working environment, regular audits. These are clear signs of a work culture that is caring and concerned.

Leaders believe that the aforesaid along with the earlier findings mentioned in the previous sections will not only motivate employees to be more productive but also lower down the attrition rate.

1.7 CONCLUSION

The HR marketplace is poised for significant growth in the next 10 years. Companies around the world are increasing their investment in HR infrastructure. The first 15 years of the 21st century have witnessed a rapidly changing world of work. The VUCA (Volatile, Uncertain, Complex, and Ambiguous) political, economic, social, and technological (PEST) environment is also evolving very fast and impacting business and managerial perspectives. As an important component of business, HRM function is under pressure and must be prepared to deal with the effects of such change. It has to understand the implications of globalization, competition, changing skill requirements, corporate downsizing, reengineering and needs to be agile enough to come up with innovative practices to bring about continuous improvement in the HRM systems and enhance employee engagement to make organizations future ready. All this has given rise to challenges for HR to tackle issues like diversity management, competency mapping, virtual teams, real time delivery models, temporary versus permanent employees, work–life balance, lean structure, knowledge management, multiple generations in work force (X, Y and the new in line—Z; with difference in their respective thought processes relating to instant gratification, autonomy, live in the present attitude, earn now and spend now, and so forth). Rigorous working is simultaneously leading to early burnout. It is difficult for an organization to have one set of policies across the board. A lot of rethinking and overhauling of existing practices is needed.

Organizations need to map their policies with requirements of future set of employees. As changes are both external as well as internal, the task of leaders and top management has further got complicated. Manufacturing sector is a very traditional sector but the research on a few successful firms brings out that they tend to have in common their sustained advantage; they rely not on technology, patents, or strategic position, but on how they manage their workforce. What is important to recognize is how success can be sustained and cannot readily be imitated by competitors. There are two fundamental reasons. First, the success that comes from managing people effectively is often not as visible or transparent as to its source. We can see a computerized information system, a particular semiconductor, a numerically controlled machine tool. Culture, how people are managed, and the effects of this on their behavior and skills are sometimes seen as the "soft" side of business, occasionally dismissed. Even when they are not dismissed, it is often hard to comprehend the dynamics of a particular company and how it operates because the way people are managed often fits together in a system. It is easy to copy one thing but much more difficult to copy numerous things.

Achieving some competitive advantage through the workforce is that it inevitably takes time to accomplish. It is almost inconceivable that a firm facing immediate short-term pressure would embark on activities that are apparently necessary to achieve some competitive advantage through people. This provides one explanation for the limited diffusion of these practices. If the organization is doing well, it may feel no need to worry about its competitive position. By the same token, if the organization is in financial distress, the immediate pressures may be too severe to embark on activities that provide productivity and profit advantages, but only after a longer and unknown period of time. Things that are measured get talked about, and things that are not don't.

It is no accident that companies in a world in which financial results are measured, a failure to measure HR policy and practice implementation dooms this to second-class status, oversight, neglect, and potential failure. The feedback from the measurements is essential to refine and further develop implementation ideas as well as to learn how well the practices are actually achieving their intended results. Most simply put, it is hard to get somewhere if you don't know where you are going. In a similar fashion, practices adopted without a deeper understanding of what they represent and why they are important to the organization may not add up to much, may be unable to survive internal or external problems, and are likely to produce less than stellar results.

After observing and analyzing the above mentioned data we can conclude that the leaders of the industry are trying to guarantee minimal attrition.

The best workplaces continue to find innovative ways to create personal and meaningful connections with their employees. The pace of change requires that company leaders seek new strategies to maintain employees. The focus on organizational culture and human capital grows, even as major economies shift to service and knowledge industries. However, the nature of work is changing in ways that make cultivating a strong personal connection difficult. Technology has created virtual workplaces. As a result, coworkers are now more widely dispersed and working at different times of the day. The mixture of people in the workforce is evolving too, spanning generations and geographies with a demographic more diverse and reflective of the communities where companies are located. Despite the global recession, employees continue to enjoy more choice of where to work. They no longer plan to stay with a company for life, leading to increased job-hopping and cross-fertilization of workplace cultures. In response to these challenges, the best workplaces focus not just on workers' basic economic and security needs, but on creating meaningful work and supportive social networks for employees. Many of the best workplaces take steps to show an individual how his work directly ties with the company's strategy and purpose.

Though no employer can guarantee that they will always have people with high levels of volition but what is in one's hands is to create an atmosphere where they can at least try to give external motivation to engage the people and facilitate intrinsic motivation.

The HR activities include development of talent pool, promote demonstrative work culture, and promote job rotation and high level of empowerment. Thus, Personal learning and development opportunities to employees are some of the reasons for staying with a company for long.

If we look at his study from the VRIO model (Barney, 1991) perspective, this paper helps to understand how an organization can offer support to an employee who adds value, possesses or develops a rare skill set so that it is almost inimitable.

Employees are not working in silos or in water tight compartments and hence the companies facilitate a smooth flow of individual from one job role to the others.

Thus, from the entry of an employee through recruitment and selection to the final exit, HR interventions are a never ending source of managing talent.

An engagement survey reflects that all this helps in reducing absenteeism and anxiety and help in providing a stress free work–life balance.

Some of the key areas of responsibility are equal treatment, equal opportunity along with diversity. Investment in personal development can continue to build retention and advancement of women.

Focus on the right environment: physical safety, psychological safety, unbiased, ethics committee, zero tolerance for harassment, fraud, respect inclusion; core values are taken very seriously; they are a part of parameters for assessment and people are recognized for it. As a result, retention is very high.

The whole environment is an enabler and is good for employees' psychological safety. These companies have had a fairly high retention rate and it is not unusual to meet employees, staff and workers alike that have been with the company for decades. Many view the company as part of their family where their grandparents, parents, siblings or spouse have worked or are working in the same company.

1.8 LIMITATIONS AND FUTURE SCOPE

This is a qualitative research providing in-depth understanding of five companies in the manufacturing sector in Pune. Time, money, human resource and availability of HR personnel were the most obvious constraints. The results of the study could be further quantitatively studied before making generalization.

Several authors have established the link between HR practices and the firm level strategies. There are, however, divided opinions about the adoption of universalistic approach to strategic HR (adoption of HR Best Practices) or contingency approach. Most of the Indian manufacturing companies are still not mature in their human resource practices. They are still practicing the mantra of one-size-fits-all, which is inappropriate given the current competitive pressures. It is essential for firms to realize and quantify the impact that effective human resource practices can have on their operational and financial performance.

The chapter details the various new age HR initiatives driven by few of the best manufacturing companies to address new age employees issues. Taking a cue from this, other companies from the same industry and organizations across sectors can review their HR efforts. We can also take the results and study if a quantifiable correlation between HR policies and organization's performance in terms of production, profit; employee's perspective like satisfaction, motivation, retention; and customer satisfaction, and so forth exists. This might help in generalization of the findings and establishing importance of transformations in the area of HR.

The limitations of this study constrain the interpretation of the findings and point to several issues for future research. Gaining a clearer understanding of the relationships between HR systems and strategy will require a longitudinal analysis. Also, the study does not include the performance of these companies and strictly goes by the literature and findings through primary data on the topic available. Therefore, testing the performance of the companies in addition to their HR practice-corporate strategy link can be an extension of this study.

> *In the end it is not the strongest of the species that survives, or the most intelligent that survives. It is the one that is most adaptable to change.*
> *−Darwin*

1.9   ANNEXURE

INTERVIEW SCHEDULE

Name of the organization:

Person Interviewed:

Designation:

Date:

Type of interview: Telephonic/Face to face/Self-Administered

As the business environment in India is undergoing unprecedented changes, the human resource function of organizations also needs to keep pace with the transformations taking place in the external environment. In the context of these changes we are conducting a research titled **"GEARING UP FOR THE FUTURE: HR TRANSFORMATIONS IN THE MANUFACTURING SECTOR"** and request your valuable time to answer our few queries. We shall be highly grateful if you could send in the filled form at the earliest to any of the e-mail IDs: manishasaxena75@ homail.com/manisha.saxena@ibsindia.org/sknandi2011@gmail.com/ subrato.nandi@ibsindia.org
Thank you for your cooperation.

1. What are the challenges faced by your organization today in the global environment in general and in India in particular?

2. The challenges faced by your organization today in managing human resources.
3. What are the transformations taking place in your industry specific sectors (in India and globally).
4. The strategies that your organization is adopting to manage human resources.
5. Key policies adopted in the areas of recruitment, selection, retention of employees, performance measurement systems, employee engagement, etc.
6. The changes which have taken place in employee-related issues in the organization over the last 5–10 years, specifically post 2008–2009 financial crisis (e.g., multiple generations in work force X, Y, Z).
7. Some best HR practices or Benchmarking HR practices you could comment on (e.g., leadership/culture)
8. Any other information you would like to furnish.

KEYWORDS

- **VUCA**
- **HR practices**
- **future organizations**
- **manufacturing**

REFERENCES

Agarwala, T. Innovative Human Resource Practices and Organizational Commitment: An Empirical Investigation. *Int. J. Hum. Res. Man.* **2003,** *14*(2), 175–197.

Baran, B. E.; Shanock, L. R.; Miller, L. R. Advancing Organizational Support Theory into the Twenty-First Century World of Work. *J. Business Psychol.* **2012,** *27*(2), 123–147.

Barney, J. B. Firm Resources and Sustained Competitive Advantage. *J. Manage.* **1991,** *17*, 99–120.

Beatty, R. W.; Schneirer, C. E. New HR roles to Impact Organizational Performance: From "partners" to "Players". *Hum. Resour. Manage.* **1997,** *36*, 29–38.

Bird, J. Work-Life Balance: Doing it Right and Avoiding the Pitfalls. *Employment Relations Today* **2006,** *33*(3), 21–30.

Boselie, P.; Dietz, G.; Boon, C. Commonalities and Contradictions in HRM and Performance Research. *Hum. Resour. Manage. J.* **2005,** *15*(3), 67–94.

Boxall, P. The Strategic HRM Debate and the Resource-Based View of the Firm. *Hum. Resour. Manage.* **1996,** *6*(3), 59–74.

Branham, L.; Hirschfeld, M. *Re-Engage: How America's Best Places To Work Inspire Extra Effort in Extraordinary Times;* McGraw Hill: Chicago, US, 2010.

Budhwar, P. S.; Khatri, N. A Comparative Study of HR Practices in Britain and India. *Int. J. Hum. Resour. Manage.* **2001,** *12*(5), 800–826.

Burke, R. J.; Ng, E. The Changing Nature of Work and Organizations: Implications for Human Resource Management. *Hum. Resour. Manage. Rev.* **2006,** *16*, 86–94.

deLeede, J.; Looise, J. K. Innovation and HRM: Towards an Integrated Framework. *Creativity Innovation Manage.* **2005,** *14*(2), 108–117.

Delery, J.; Doty, D. H. Modes of Theorizing in Strategic Human Resource Management: Tests of Universalistic, Contingency and Configurational Performance Predictions. *Academy Manage. J.* **1996,** *39*(4), 802–835.

Felstead, A.; Jewson, N.; Phizacklea, A.; Walters, S. Opportunities to Work at Home in the Context of Work-Life Balance. *Hum. Resour. Manage. J.* **2002,** *12*(1), 54–76.

Greene, B. One Workforce, Many User Experiences: Delivering HR Service to "All those Gens!". *Workforce Solutions Review*, pp 24–27, (December 2013/January 2014).

Gunavathy, J. S. Work-Life Balance Interventions Prevalent in the Indian Industry. *South Asian J. Manage.* **2011,** *18*(2), 108–127.

Krishnaveni, R.; Sripirabaa, B. Capacity Building Process for HR Excellence. *VISION—J. Business Perspect.* **2008,** *12*(2), 1–13.

Lepak, D. P.; Bartol, K. M.; Erhardt, N. L. A contingency Framework for the Delivery of HR Practices . *Hum. Resour. Manage. Rev.* **2005,** *15*, 139–159.

Levering, R.; Erb, M. Emerging Trends in People Management. *Swiss Business* **2011,** pp 31–32.

Manimala, M. J.; Kumar, S. Training Needs of Small and Medium Enterprises: Findings from an Empirical Investigation. *IIM Kozhikode Soc. Manage. Rev.* **2012,** *1*(2), 97–110.

Marshall, C.; Rossman, G. B. *Designing Qualitative Research,* 4th ed.; Sage Publications, Inc.: Thousand Oaks, California, USA, 2006.

Martin-Alcazar, F.; Romero-Fernandez, P. M.; Sanchez-Gardey, G. Transforming Human Resource Management Systems to Cope with Diversity. *J. Business Ethics* **2012,** *107*(4), 511–531.

Mesch, D. J. Management of Human Resources in 2020: The Outlook for Nonprofit Organizations, 70, *Public Administration Review, Supplement to Volume 70 The Future of Public Administration in 2020,* **2010,** S173–S174.

Nishii, L. H.; Lepak, D. P.; Schneide, B. Employee Attributions of the "Why" of HR Practices: Their Effects on Employee Attitudes and Behaviors, and Customer Satisfaction. *Pers. Psychol.* **2008,** *61*, 503–545.

Osterman, P. Choice of Employment Systems in Internal Labor Markets. *Ind. Relat.* **1987,** *26*(1), 46–67.

Pfeffer, J. Producing Sustainable Competitive Advantage Through the Effective Management of People. *Academy Manage. Exec.* **1995,** *9*(1), 55–69.

Pocock, B. Work-Life 'Balance' in Australia: Limited Progress, Dim Prospects. *Asia Pac. J. Hum. Resour.* **2005,** *43*(2), 198–209.

Porter, M. E. The Five Competitive Forces That shape Strategies. *Harvard Business Rev.* **2008,** *86*(1), 78–93.

Prahalad, C. K. Developing Strategic Capability:An Agenda for Top Management. *Hum. Resour. Manage.* **1983,** *22*(3), 237–254.

Rhoades, L.; Eisenberger, R. Perceived Organizational Support: A Review of the Literature. *J. Appl. Psychol.* **2002,** *87*, 698–714.

Stockdale, M.; Crosby, F. *The psychology and management of workplace diversity.* Blackwell Publishing: Malden, MA, 2004.

Stroh, K. L.; Caliguiri, P. M. Increasing Global Competitiveness through Effective People Management. *J. World Business* **1998,** *33*(1), 1–16.

Syed, J.; Kramar, R. Socially Responsible Diversity Management. *J. Manage. Organ.* **2009,** *5*(5), 639–651.

Thompson, C. A. Managing the Work-Life Balancing Act: An Introductory Exercise. *J. Manage. Educ.* **2002,** *26*(2), 205–220.

Tomar, A. Effect of Organizational Strategy on Universalistic or Contingent HR Practices in Indian Manufacturing. *Indian J. Ind. Relat.* **2011,** *47*(2), 306–320.

Ulrich, D. Measuring Human Resources: An Overview of Practice and a Prescription for Results. *Hum. Resour. Manage.* **1997,** *36*(3), 303–320.

Ulrich, D.; Wiersema, M. F. Gaining Strategic and Organizational Capability in a Turbulent Business Environment. *Academy Manage. Exec.* **1987–1989,** *3*(2), 115–122.

Wright, P. M.; McMahan, G. D.; McWilliams, A. Human Resources and Sustained Competitive Advantage: a Resource-Based Perspective. *Int. J. Hum. Resour. Manage.* **1994,** *5*(2), 301–326.

CHAPTER 2

FACTORS AFFECTING SELF-SERVICE BANKING ADOPTION IN FEMALES: A DISCRIMINANT ANALYSIS

PURVA KANSAL

University Business School, Panjab University, Chandigarh, India, E-mail: purvakansal@pu.ac.in

CONTENTS

ABSTRACT

The future organizations will be full of technology usage for customers and as females are increasingly becoming important customers using banking services, this chapter is an attempt to understand the factors, which influence adoption of technology among females. Marketers need to understand these differences more specifically when talking in terms of services because inseparability means the demographics and consumer personality is part of the equation. Therefore, understanding the gender differences in usage and adoption of services will give marketers a necessary edge. It is this aspect of services which was focus of the current study. To test the proposed hypotheses, a survey was done and data were collected from 125 respondents. The sampling frame was defined as 26 tier II cities of India. Out of these 26 cities, 4 cities were chosen for data collection, that is, Chandigarh, Ahmedabad, Dehradun, and Ludhiana. Discriminant analysis (stepwise) was used to determine which factors act as strong discriminators among females for intention to use self-service banking. The results of the study indicate that the marketing strategist should concentrate on increasing self-efficacy perception of the female segment and at reducing perception of perceived risk of the female customers to improve adoption of technology or self-service banking by the female population of the country. These two variables are required to change the perception of perceived ease of use and perceived usefulness, two very important variables for the female segment. The results of the study stress on relative and customerization-based strategies. Therefore, indicated that the four were significant discriminators of intention to use and should be invested on more by strategists to increase the adoption rate and ensure inclusion of the female gender in the self-banking services.

2.1   INTRODUCTION

Customerization has been the solution given by service industry to increased competition and imitability in service industry. Customerization means giving transactional control to the customer for every transaction and every item in the transaction. Scope of customerization as a strategy is more than that of customization, where the consumer is allowed to customize a particular product at an additional cost; however customerization allows the consumer control of every transaction. As a strategy, transference of control to the consumer is interpreted as more perceived power by the consumer. Only way companies can achieve this degree of flexibility in processes is with help of technology. Earlier uses of technology were in the form of employees using technology to facilitate the delivery process, however; these days consumers are progressively using technology to serve themselves. This infusion of technology in the service blueprint, as a self-service component has altered the very essence of service industry which traditionally competed on the basis of personal contact and relationship management. This change is especially true in information-processing industries such as banking where the companies have proactively used technology in their service blueprints so as to provide customerization and transfer control to their customers; be it in the form of ATM, online banking, or mobile banking. These services that allow customerization require higher degrees of involvement from the customers and are often referred to as self-service technologies or banking.

Customerization as a strategy, in a company, is a beneficial option because it allows the company to reap benefits like reduced cost of operations, increased efficiency of service processes, and improved quality (Legris et al., 2003; Meuter et al., 2000). However, all the benefits are function of adoption rate of the introduced technology-based service by the customers, for example, if the service is not accepted by the customers, then reaping the benefits of the strategy become a distant dream for the company. According to the popular literature, consumers are often reluctant to adopt self-servicing technologies because it requires significant behavioral change. In the context of self-service banking technologies this behavioral change requires the consumer to become coproducers of service and assume the responsibility for delivery of service and therefore, satisfaction becomes partly theirs (Bendapudi and Leone, 2003; Meuter et al., 2005). As a resultant, it is seen that as consumers are reluctant to adopt self-service technologies the companies are slow to realize rate of return on their very expensive self-service banking technologies.

To help companies break this cycle and influence this adoption rate of self-service banking technologies, researchers in the past have suggested a model-based approach. In this model-based approach, many theories have developed philosophizing the models, for example, the technology acceptance model (TAM)(Davis, 1989; Edgett and Parkinson, 1993; Zeithaml et al., 1985), the theory of reasoned action (Fishbein, 1979), task–technology fit model (Goodhue and Thompson, 1995), and innovation diffusion theory (Rogers, 2010). TAM: proposed by Davis in 1989, TAM posits that behavioral intention is determinant of actual use and behavioral intention is in turn determined by perceived ease of use and perceived utility. Model states that perceived utility is the user's perception of the extent to which using technology would enhance their performance, and perceived ease of use is perception of the user of the amount of effort required to use the technology.

The theory of reasoned action: proposed by Fishbein in 1979, this particular model had received considerable attention in 1980. This model is used to predict behavioral intentions of the consumer. The theory posits that a consumer's behavioral intentions are antecedents to behavior and are functions of beliefs about the likelihood of a particular behavioral outcome. The market has further bifurcated beliefs into two sets, that is, behavioral and normative.

Task technology fit model (TTF) is a model that overlaps with TEM significantly. Task technology fit model focuses on individual performance attribute able to actually use while TAM focuses on the intention to use. Therefore, this model is more cured around ability of IT to support the task, that is, it is centered around matching the capabilities of the technology to the demands of the task the consumer is supposed to perform.

The innovation diffusion theory (IDT) is a theory which concentrates significantly on innovation characteristics such as relative advantage, compatibility, complexity, trialability, and observability. The theory suggests that these characteristics can be used by a particular company to explain the adoption of innovations and a consumer's decision-making process.

Though these models are all significantly different yet a brief analysis indicated that theory of reasoned action, task technology fit, and innovation diffusion theory in some way was similar to TAM as well. Such as the relative advantage construct in innovation diffusion theory was similar to perceived usefulness in the TAM and complexity construct was similar to perceived ease of use (Lee et al., 2011). Similarly, TTF was more popular in terms of finding fit of technology and the demands of the task. Therefore, TAM was found to be the most popular model for studying acceptability of technology-oriented services from customer perspective and also most appropriate in the current scenario. The popularity of the model-based approach has been so popular that a general search on TAM-based studies on Google Scholar in May 2016 resulted in more than 1,970,000 hits in 0.02 s with the phrase being either in their title or content.

Armed with the knowledge gained from these models, companies have been developing strategies to promote self-service banking. However, in India, these strategies have not been able to help banks to reach its true potential in terms of self-service banking technologies. Statistics indicates that self-service banking is suffering from a low adoption rate, that is, 65% of the online banking registered customers remain inactive (public as well as private banks) and the number of registered users itself ranges between only 2 and 8% of overall number of banking transactions across channels (EY, 2014).

A search for possible explanations into this lack of adoption indicated an answer to this question could be in demographics. A review of literature (Zmund, 1979) had indicated in his studies that individual differences such as demographics, gender, age, level of education, and personality-related variables influenced the consumer adoption of technology. Many researchers have supported this argument over the years and indicated that in terms of perception regarding technology, there are significant differences across gender right from the risk dimensions to the perception of capability (Chiu et al., 2005; Elliott and Hall, 2005; Karjaluoto et al., 2010; Wang et al., 2009).

Therefore, based on the review of existing literature it can be concluded that there are significant differences across gender and ignoring these in marketing strategies forfeits the very purpose of adopting customerization strategy. It is within this backdrop that the current study has been undertaken in India for females. The current study was undertaken with a specific objective of determining the factors of the TAM; which discriminates between groups based on satisfaction within the females of the society.

The chapter is an attempt to understand the factors that influence adoption of technology among females. Marketers need to understand these differences, especially when talking in terms of services, because inseparability means the demographics and consumer personality is part of the equation. Therefore, understanding the gender differences in usage and adoption of services could give marketers a necessary edge. It is this aspect of services that has been under researched and is the focus of the current study. The research would help add to the existing base of literature on TAM and technology-based services and also would help practitioners develop understanding and strategies for increasing the adoption rate among female customers.

2.2 REVIEW OF LITERATURE

In order to study the factors that influence adoption of technology among females, a review of literature from leading journals and databases was conducted. The review has been organized in six sections based on the dependent and independent variables of the study.

2.2.1   THEORY DEVELOPMENT: TECHNOLOGY ACCEPTANCE MODEL

Technology acceptance model was proposed by Davis (Davis, 1993). The model proposes that the attitude of a person towards technology is determinant of actual use and the attitude is determined by two factors: perceived ease of use and perceived utility. The model postulates these two variables: perceived ease of use and perceived usefulness as mediating variables for a set of external variables. Over the years, many researchers have tested and proven the model strength across different technology-oriented products. There has been a lot of evidence for the effect of perceived usefulness and perceived ease of use on attitude and has been provided by multiple authors over the years (Chau, 1996; King and He, 2006; Mun and Hwang, 2003; Venkatesh and Davis, 1996; Vijayasarathy, 2004). Extending this model and studying it for moderating effects, some researchers have indicated that the influence of perceived ease of use and perceived usefulness on attitude will be moderated by gender (Elliott and Hall, 2005; Karjaluoto et al., 2010; Wang et al., 2009). Therefore, based on the findings of past researchers, it is hypothesized that perceived ease of use and perceived utility would vary across three groups of intention to use for females that is use, no usage, and neutral.

The choices of factors, which influence the satisfaction from self-service banking, were shortlisted on the basis of the review of literature that was based on TAM. Though limited work was found on females specifically, there was rich literature on gender differences and factors which generically influence the technology adoption rate. These are the studies which were used in theory development of current study.

2.2.2   SOCIAL NORM

This set of items is defined as a person's perception that most people who are important to him think he should or should not perform the behavior in question (Fishbein, 1979; Fishbein and Ajzen, 1975). Past research indicates that women are more people- and relationship-oriented than men and therefore, stronger determinant of behavior is seen in women than in men (Venkatesh et al., 2000). This relationship orientation motivates females to stay within the social norms and to maintain their relationships. It is due to this reason that females give relatively more weightage to social norms and are quicker to adopt something that is required as a social norm. Therefore, it is proposed that social norms would influence the purchase decision and, thus, act as a discriminator between groups based on three groups of intention to use for females.

2.2.3   PROPOSITION: PURCHASE INTENTION WOULD BE SIGNIFICANTLY INFLUENCED BY SOCIAL NORMS

2.2.3.1   PERCEIVED RISK

Perceived risk as a construct refers to uncertainty regarding expected benefits from a product or service (Bauer, 1960). It is the perception of negative outcome which influences the adoption of a technology. To test this perception many scholars have extended TAM to include perceived risk as an external variable (Featherman and Pavlou, 2003; Kesharwani and Bisht, 2012; Lee, 2009; Yiu et al., 2007). Researchers have indicated that women have greater perceived likelihood of negative outcomes and a lesser expectation of enjoyment which is partially mediated by their lower propensity toward risky choices in gambling, recreation, and health domains (Harris et al., 2006). Furthermore, researchers have indicated that there is a presence of socially instilled belief that risk-taking is a highly valued masculine tendency and, therefore, it motivates high levels of risk-taking across contexts in men (Byrnes et al., 1999). Based on these arguments, it is postulated that because of the social beliefs, especially in a patriarchal society like India, females will be inclined to be less risk-taking and their behavior and decision-making process will be influenced by their perception of risk. Therefore, their intention to use will be influenced by their perceived risk.

2.2.4   PROPOSITION: PURCHASE INTENTION WOULD BE SIGNIFICANTLY INFLUENCED BY PERCEIVED RISK

2.2.4.1   SELF-EFFICACY/CAPABILITY

Most of the self-service banking methods require the use of technology; therefore, capability was included as an external variable in the study.

Social learning theory states that psychological procedures alter expectations of personal efficacy (Bandura, 1977, p. 79). Therefore, as the psychological procedures for males and females are different, their personal efficacy perceptions are also different. This has been supported by other researchers who in their gender-based studies, in the context of computer literacy, have indicated that women have more computer and internet-related anxiety and men had higher self-efficacy than women (Durndell and Haag, 2002; Ong and Lai, 2006; Schwarzer et al., 1997). Therefore, on the basis of these arguments, it is propositioned that self-efficacy would influence the intention to use in females.

2.2.5 PROPOSITION: PURCHASE INTENTION WOULD BE SIGNIFICANTLY INFLUENCED BY SELF-EFFICACY

2.2.5.1 TECHNOLOGY DISCOMFORT

Technology discomfort is referred to as the tendency of an individual to be uneasy, apprehensive, stressed, or have anxious feelings about the use of technology (Venkatesh, 2000). In the context of gender, it was found that, in general, women exhibit a higher discomfort with computer-based technologies as compared to men (Durndell and Haag, 2002; Gefen and Straub, 1997; Ong and Lai, 2006). Based on these arguments, the current study proposes that when it comes to deciding whether to use a technology-based service or not the female segment would experience more discomfort and anxiety and would be more inclined to choose not to use the technology-based service.

2.2.6 PROPOSITION: PURCHASE INTENTION WOULD BE SIGNIFICANTLY INFLUENCED BY TECHNOLOGY DISCOMFORT

2.2.6.1 CONCEPTUAL MODEL

The current study is an attempt to understand the factors that influence the adoption of technology among females in TAM. Based on the review of literature following conceptual model has been proposed for the current study. The conceptual model tries to study the relationship between perceived ease of use, perceived utility, and intention to use. This part of the model is as per technology acceptance more. Further, based on the review of literature a set of independent variables were used in order to understand the effect of these variables on intention to use. These independent variables were technology discomfort, perceived risk, social norm, and self-efficacy. Though the model indicates perceived ease of use and perceived utility as mediating variables yet the primes of the current study was limited to understanding the influence of four factors ,that is, technology discomfort, perceived risk, social norm, and self-efficacy on intention to use when perceived ease of use and perceived utility are also in the relationship. Due to this reason stepwise discriminant analysis was chosen for analysis.

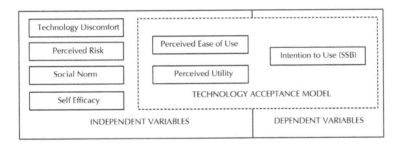

In order to meet the objective of the current study, based on the conceptual model, following hypotheses are being proposed for testing.

- H1: Technology discomfort has a significant discriminating power across intention to use self-service banking in females with perceived ease of use and perceived utility.
- H2: Perceived risk has a significant discriminating power across intention to use self-service banking in females with perceived ease of use and perceived utility.
- H3: Social norm has a significant discriminating power across intention to use self-service banking in females with perceived ease of use and perceived utility.
- H4: Self-efficacy has a significant discriminating power across intention to use self-service banking in females with perceived ease of use and perceived utility.

2.3   RESEARCH METHODOLOGY

2.3.1   RESEARCH DESIGN

An exploratory as well as descriptive research design was adopted to test the hypothesis. Exploratory research design was adopted where an exhaustive review was conducted of the existing literature in order to develop the conceptual model and propositions for the study on which the hypotheses have been based. A structured non-disguised questionnaire was used to collect data.

2.3.2   SAMPLING

The unit of analysis for the current study was females who had experienced service failure. In order to collect data from the unit of analysis a multi-layered sampling plan was developed. To collect data for testing off the proposed hypothesis sampling unit was decided as tier II cities of India in the first stage. In order to narrow down the tier II cities, the something frame was defined as survey was done and data were collected from 125 respondents. The sampling frame was defined on basis of the list issued by the Government of India to allot house rent allowance (HRA). This list had 26 tier II cities. In this stage, four cities were chosen for data collection, that is, Chandigarh, Ahmedabad, Dehradun, and Ludhiana on the basis of random sampling using random tables. The sampling unit in stage II was defined as individuals across Chandigarh, Ahmedabad, Dehradun, and Ludhiana. In order to collect data from these individuals snow ball sampling was used. In the second stage, snowball sampling methodology was used in order to reduce the effect of socially desirable responses amongst respondents.

2.3.3   DATA COLLECTION INSTRUMENT

Data was collected using a structured non-disguised questionnaire. The sections of the questionnaire were defined on basis of the conceptual model variables and, therefore had eight sections. For each one of the variables in the study and standardized constructs were adopted to quantify the responses, that is, perceived usefulness and perceived ease of use (Davis, 1985); technology discomfort (Parasuraman, 2000); social norm

(Taylor and Todd, 1995); perceived risk (Parasuraman, 2000); and self-efficacy (Chen et al., 2001); the items for attitude towards intention to use were developed. The last section of the questionnaire was regarding demographics of the respondents.

2.3.4   CLEANING AND EDITING OF DATA

The collected data was cleaned before any sort of analysis was to be performed on the data. This step was undertaken in order to detect and remove any errors and inconsistencies in the data set due to incomplete or inaccurate data. In this particular stage, in the present study the data was analyzed for incorrect entries, outliers, and missing values. Preliminary analysis indicated that out of 125 questionnaires 124 had provided responses for education degree, 125 had responded for educational qualification, and 118 had responded for income. However, the information about independent and dependent variables was complete in the whole questionnaire, and an analysis of the variance indicated that the responses were involved responses. Therefore decision was taken to retain all 125 questionnaires and not to discard any. Therefore, the sample size of the current study warns 125 female respondents.

2.3.5   PSYCHOMETRIC PROPERTIES OF THE SCALE

Face validity of the questionnaire was tested through pilot testing. The survey instrument was written in English and was pretested on a small sample of 20 respondents. Face and content validity of the instrument and its items were concluded by various researchers with experience in conducting surveys. The internal consistency and reliability of the scale was measured using Cronbach coefficient alpha. According to works of Nunnally (1978), for purpose of the basic research, a Cronbach alpha of 0.70 or higher is sufficient. Cronbach alpha for the adopted scale met this limit, and it was found that a Cronbach alpha value for all constructs was higher than 0.70.

The data analysis to test the hypothesis of the study was done through discriminant analysis. In order to test the mediating variables of the study, a stepwise method for discriminant analysis was adopted. In the first step perceived ease of use was added, in the second perceived usefulness, and later the other variables. The methodology help evaluate the order of importance of the variable.

2.3.5.1   PRESENCE VS MEDIATING EFFECT

Technology acceptance model uses perceived ease of use and perceived useful-ness as mediating variables. However, for the current study though the presence of both perceived ease of use and perceived usefulness have been acknowl-edged, they are not being truly treated as mediating variables. The objective of the study was to help in strategic decision-making across the segments of intention to use for female respondents. To meet this objective the statistical analysis required discriminant analysis. However, the methodological construct discussed in the preceding paragraph indicates that though to some extent inter-pretation can be made for mediation, there are certain limitations and, there-fore, instead of mediation only presence of the variables is being considered.

Barron and Kenny's approach for the indirect effect is inferred by testing a set of hypothesis associated with a, b, and c paths (Baron and Kenny, 1986). This model assumes a three-variable system such that there are two causal paths feeding into the outcome variable: the direct impact of the independent variable (Path c) and the impact of the mediator (Path b). There is also a path from the independent variable to the mediator (Path a). Judd and Kenny (1981) suggested implementing a process-based approach by testing only path b and c because they postulated that mediates treatment effects should be an important part of most evaluation studies (Judd and Kenny, 1981). Though a process-based approach has been used in the discriminant analysis to test these two paths by using stepwise method, yet the method is not testing true mediation. It is important to note here that the study is not testing mediation effect, it is only testing the influence of four factors, that is, technology discomfort, perceived risk, social norm, and self-efficacy on intention to use when perceived ease of use and perceived utility are also in the relationship. It was due to this reason that stepwise discriminant analysis was chosen for analysis.

2.4   FINDINGS AND DISCUSSION

A descriptive analysis of the data indicated that out of 125 female respon-dents, 42.7% were graduates, 40.8% were post graduates and only 10%

were undergraduates. A very small percentage of respondents belonged to doctorate and professional diploma categories. The age of the respondents indicated that 30.2% of the respondents were in their 20s, 36.4% were in their 30s while around 35% of the respondents were in their 40s and 50s. The sample seemed to be almost equally balanced among science and non-science students, that is, 40, 42%, respectively. Most of the population seemed to be working with some sort of the profession with only 32% being house wives. An analysis of the screening question indicated that a total of 48% of the respondents said that they were familiar with self-service banking technologies for 3–8 years while 34% said they were familiar or occasionally used the self-service banking technology for less than 2 years while 10% said they have never used a self-service banking service. Therefore, it was concluded that the sample was diverse enough for analysis.

A descriptive statistics were generated for independent variables of the study in order to understand the nuances of the respondents. The statistics indicated that 62.4% of the females had a positive attitude towards the self-service banking technologies, while 25.6% of the female respondents were neutral in terms of their attitude towards self-service banking technologies and 12% had a negative attitude towards self-service banking technologies. Majority of the respondents, that is 47%, though were skewed towards positive side of using self-service banking technologies rather than cash yet 30% of the respondents were of the opinion that using cash was safer than using self-service banking technologies, and around 23% were on the fence and undecided about self-service banking technologies. It was felt that converting these 23% of the respondents into a positive mind frame would be easier for the companies as compared to 30% of the respondents carried negative attitude. An analysis of the mean values of intention to use variable indicated that 45% of the respondents were using or had any intention to use self-service banking technologies. If clubbed together the attitude and intention to use indicated that 30% of the respondents who were on the fence in terms of their attitude were also opting not to use self-service banking technologies. Therefore, it could be argued that attitude needed to be changed before companies could think in terms of getting a higher intention to use. The attitude towards studied through two variables, i.e., perceived ease of use and perceived usefulness. And exploratory analysis of these two variables indicated that the majority of the female respondents, in the current study, considered self-service banking technologies as useful. 52% of the respondents were in agreement that the SSB technologies were useful while only 14% of the respondents were of the opinion that they were not. As seen before, a good chunk of respondents, that is 34%, of the female respondents were on the fence with their attitude regarding perceived usefulness of self-service banking technologies. A similar trend was seen in exploratory statistics of perceived ease of use. Majority of the respondents, that is 55%, perceived it easy to use self-service banking technologies while 20% perceived it to be a difficult proposition. In line with the trend, again 25% of the respondents were on the fence regarding their perception toward ease of using self-service banking technologies. A plausible explanation for this attitude was found in the technology discomfort variable. The descriptors of technology discomfort indicated that the mean value of this variable was 2.3 and 54% of the respondents were uncomfortable with the use of self-service banking technologies. Only 3% of the respondents were very comfortable in using self-service banking technologies. A similar trend was seen in the descriptors of perceived risk variable. The average value for perceived risk was 3.1. This value was a little better than that of technology discomfort. However, 62% of the respondents were on the fence, i.e., neutral about the perceived risk of self-service banking technologies. The descriptive statistics generated for the study indicated that lower indication among the sample for intention to use could be attributed to perceived ease of use and perceived usefulness which in turn were influenced by the negative perception of the technology and perceived risk. To analyze this, further of correlation analysis was undertaken for the abovementioned variables before testing the hypothesis for discriminant analysis.

Correlation analysis indicated that there was support for relationship between intention to use and perceived ease of use, and perceived usefulness (Table 2.1). The statistics indicated that the argument, that in order to influence the intention to use for self-service banking technologies companies should influence female customer's perceived ease of use and perceived usefulness, could be supported. The results indicated that there was a significant and positive relationship between intention to use self-service banking technologies and perceived ease of use and perceived usefulness. Therefore, if a company could increase the female consumers' perception that self-service banking technologies were easy to use and put a more useful than using cash, then the intention to use these technologies would increase.

TABLE 2.1 Correlation Analysis.

		Perceived ease of use	Perceived usefulness		
Intention to use	Pearson Correlation	0.516**	0.557**		
	Sig. (2-tailed)	0.000	0.000		
		Technology discomfort	Social norm	Self-efficacy	Perceived risk
Perceived ease of use	Pearson Correlation	−0.681**	0.013	0.826**	−0.151
	Sig. (2-tailed)	0.000	0.886	0.000	0.093
Perceived usefulness	Pearson Correlation	−0.673**	0.124	0.727**	−0.114
	Sig. (2-tailed)	0.000	0.168	0.000	0.205

**Correlation is significant at the 0.01 level (2-tailed).

A further correlation analysis indicated that there was significant relationship between self-efficacy and technology discomfort. The results indicated that perceived ease of use and perceived usefulness were positively and significantly correlated to self-efficacy, that is, an individual's belief in one's capability to handle computer-related devices. The results further indicated a negative relationship between technology discomfort and perceived ease of use, and perceived usefulness. Interpreting it together these two results indicate that if a particular respondent was uncomfortable with technology and had little self-efficacy then there would have low perception of ease of use and usefulness. Therefore, the intention to use self-service banking technologies for such respondents was found to be lower. Preliminary results of correlation analysis indicated that out of all the independent variables which were considered for the current study only technology discomfort and self-efficacy were the major external variables influencing attitude offered respondent. Therefore, based on these preliminary results it could be argued that if a company concentrated on reducing a customer's technology discomfort and increasing self-efficacy the company could increase the rate of intention to use significantly. To find support for this and to get a more directional strategic orientation, discriminant analysis was used on the current data set. The intention for analyzing data through discriminant analysis was to find the predictive and discriminating ability of the variables of the study across three groups of intention to use, i.e., who did not have intention to use SSB, currently undecided respondents and respondents who intended to use the SSB technologies.

The results of discriminant analysis indicated that there was significant variation between the factors influencing self-service banking in terms of responses of the respondents belonging to different clusters regarding intention to purchase. For discriminant analysis, purchase behavior was collapsed into three groups: 1 represented respondents who did not have intention to use SSB, 2 represented respondents who were currently undecided; and 3 represented respondents who intended to use the SSB technologies. A stepwise discriminant analysis was applied on the data set using Wilks' lambda methodology and probability of F as a basis for choosing the variable to input into the data.

Before interpreting the results of discriminant analysis, output tables to check validity of the data for discriminant analysis were analyzed and interpreted. As a first step, equality of group means test was interpreted to check for the variables excluded from the model and the variable that maximized F statistics. As a principle, minimum F value of 2.9957 is required for the entry of variables into discriminant function as variables explaining unique variation (at 0.05 level of significance) (Sharma and Chander, 2007). The F value in current analysis was below this postulated value for two variables that is perceived risk (2.756) and subjective norm (0.401). Therefore, these two variables did not qualify for entry into the discriminant analysis. Box's M tests the null hypothesis that the covariance matrices do not differ between groups formed by the dependent (Field, 2013). In order to continue the discriminant analysis, this test should not be significant so that the null hypothesis that the groups do not differ can be retained. Current data set did not meet this assumption, that is, Box's M value was 36.187 at $p = 0.032$; thereby accepting the null hypothesis of equality of covariance's (Table 2.2). Box's M test indicated that discriminant analysis could be used to test the hypothesis.

TABLE 2.2   Test of Assumptions for Discriminant Analysis.

Intention to use	Log determinant		
1 (no purchase)	−3.848	Box's M	36.187
3 (purchase)	−3.669	F	1.663
2 (neutral)	−4.666	Sig	0.032
Pooled within groups	−4.022	−	−

Box's M tests null hypothesis of equal population covariance matrices.

The output sheet for stepwise analysis indicated that there were four models, which were generated in the analysis, Model 1: perceived ease of use was entered; Model 2: perceived usefulness was entered; Model 3: self-efficacy; and Model 4: with perceived risk along with other variables. Maximum number of steps were 12; maximum significance of F to enter was 0.05 and minimum significance of F to remove was 0.10 (Table 2.3).

TABLE 2.3 Wilks' Lambda.

Step	Number of variables	Lambda	Exact F			
			Statistic	**df1**	**df2**	**Sig.**
1	1	0.422	83.622	2	122.000	0.000
2	2	0.356	40.839	4	242.000	0.000
3	3	0.322	30.515	6	240.000	0.000
4	4	0.303	24.318	8	238.000	0.000

Based on these four models, the output table of Eigenvalues indicated that in the last model two functions were generated for the data set and Wilks' lambda indicated that out of these two functions first function was significant discriminator, and it explained 93.3% variance (Table 2.4) in the data set with a chi square of 143.979 and $p<0.05$. The second function was also significant but it explained only 6.7% of the variance in the data set with a chi square of 15.424 and $p<0.05$. Therefore, it was decided to proceed with both the functions indicating that the factors which loaded on function 1 were significant discriminators of clusters based on intention to use self-service banking among females.

TABLE 2.4 Eigenvalues and Wilks' Lambda.

Function	Eigenvalues	% of variance	Cumulative %	Canonical correlation	Test of function(s)	Sig
1	1.906[a]	93.3	93.3	0.810	1 through 2	0.000
2	0.137[a]	6.7	100.0	0.347	2	0.001

[a]First two canonical discriminant functions were used in the analysis.

The structure matrix indicated that both the mediating variables, that is, perceived ease of use and perceived usefulness loaded on function 1 along with self-efficacy. Technology discomfort also loaded on this function, however, it was not included in the analysis because its F value was less than the desired minimum value of 2.9957 (at 0.05 level of significance) (Sharma and Chander, 2007). It is important to note that technology discomfort though an important indicator for intention to use self-service banking technologies was not an important discriminator across 3 groups of intention to use. Therefore, function 1 had three variables; perceived ease of use, perceived usefulness, and self-efficacy. Function 2 included two variables: perceived risk and subjective norm (Table 2.5). However, the value of F statistics was less than 2.9957 and the variable was not included in discriminant analysis.

These results indicated a support for hypothesis 4 and hypothesis 2, that is, self-efficacy has a significant discriminating power across intention to use self-service banking in females through perceived ease of use and perceived utility, and perceived risk has a significant discriminating power across intention to use self-service banking in females through perceived ease of use and perceived utility. No support was found for hypothesis 1 and hypothesis 3.

TABLE 2.5   Structure Matrix.

	Function	
	1	**2**
Perceived ease of use	0.847[a]	0.175
Perceived usefulness	0.819[a]	−0.418
Self-efficacy	0.782[a]	0.152
Technology discomfort[b]	−0.542[a]	0.269
Perceived risk	−0.057	0.535[*]
Subjective norm[b]	−0.011	−0.019[a]

[a]Largest absolute correlation between each variable and any discriminant function
[b]This variable not used in the analysis.

To further test the extent of discrimination, the function group centroid output table was analyzed (Table 2.6 and Fig. 2.1). The dependent variable was categorical in nature and had three categories: 1 for no intention to use; 2 for neutral; and 3 for intention to use. For the cluster number 1, no intention to use, the function 1 loaded at -3.037, indicating that for this cluster the perceived ease of use, perceived usefulness, and self-efficacy were all very low, that is, the female respondents in this cluster perceived the self-service banking technologies to be difficult to use, as having low usefulness and also evaluated themselves for having low self-efficacy or ability to learn how to use these technologies. These results were in concurrence with previous results which indicated that women have more computer and Internet-related anxiety and men had higher self-efficacy then women (Durndell and Haag, 2002; Ong and Lai, 2006; Schwarzer et al., 1997).

Canonical Discriminant Function Coefficients		
	Function	
	1	2
Perceived risk	.073	1.203
Perceived usefulness	.546	−1.477
Perceived ease of use	.579	1.134
Self-efficacy	.639	.401
(Constant)	−6.262	-3.770

Unstandardized coefficients.

Standardized Canonical Discriminant Function Coefficients		
	Function	
	1	2
Perceived risk	.043	.713
Perceived usefulness	.386	−1.043
Perceived ease of use	.411	.804
Self-efficacy	.434	.272

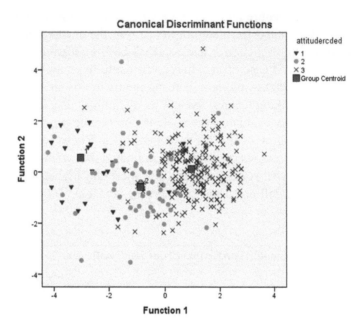

FIGURE 2.1   Discriminant analysis.

TABLE 2.6   Functions at Group Centroids

attitude	Function	
	1	**2**
1.00	−3.037	.563
2.00	−.878	−.576
3.00	.944	.128

Unstandardized canonical discriminant functions evaluated at group means.

The second function indicated that though it explained only 6.7% of the total variance in the data set, it was a significant function; indicating that perceived risk was a strong influencer/discriminator of intention to use self-service banking services in females. These results were in agreement with past research, which indicated that women perceived more risk in online purchase than men and are more risk averse than men (Garbarino and Strahilevitz, 2004; Harris et al., 2006). Therefore, given the high information-processing nature of banking industry, the perception of financial and security risk is expected to be high indicating a negative relationship between behavior and perception of risk, for example, if female respondents perceive risk to be high then their intention to use self-service banking would be less. Combining the two functions across three clusters; the results indicate that out of the four factors shortlisted, that is, technology discomfort, self-efficacy, perceived risk, and subjective norm; only self-efficacy and perceived risk were significant discriminators when perceived ease of use and perceived usefulness were present. Interestingly, the results of the study indicated that technology discomfort and subjective norm were not important discriminants for the female respondents; indicating that the gender gap was reducing in terms of these variables. This could be attributed to increased education and increased independence of the female sample and population. The demographic statistics of the sample indicated that 42% of the samples were graduates while 41% of the samples were post graduates. Majority of the female respondents were involved in some sort of a profession and only 32% being house wives. The exposure to the professional world could be a reason for the reduction in the importance of subjective norm perception in the female respondents. The professional profile could be attributed for decreased technology discomfort in the female respondents.

An analysis of Canonical Discriminant function Coefficients was done. The generic discriminant function is as follows:

$$D = V_1X_1 + V_2X_2 + V_2X_2 \ldots \ldots \ldots + a$$

Where D is discriminate function, V is the discriminant coefficient or weight for that variable, X is respondent's score for that variable, and A is constant.

Therefore, the discriminant function 1 based

$$F = -6.262 + 0.386 \, X_4 + 0.411 \, X_3 + 0.434 \, X_2$$

Function 2

$$F = -3.770 + 0.713 \, X_1$$

Where X_1 = perceived risk.
X_2 = self-efficacy
X_3 = perceived ease of use
X_4 = perceived usefulness
X_5 = technology discomfort
X_6 = social norm.

The discriminant functions are interpretted in the same manner as regression equation and therefore, the result indicates that other things constant and one unit change in the perceived usefulness of the respondent led to 0.386 units change in the intention to use self-service banking. Relatively speaking the most important variable in the cluster was self-efficacy with the discriminant coefficient of 0.434 followed by perceived ease of use and perceived usefulness.

Therefore, the results of the study indicated that a company should invest in increasing the perception of self-efficacy among its female users and that would help increase the perceived ease of use and perceived usefulness and also lead a positive intention to use. To increase the perception of self-efficacy strategies like free demos, personal contact, and encouraging trial by female customers should be encouraged. This indicates that though three were significant discriminators of intention to use that is, self-efficacy, perceived ease of use, and perceived usefulness and were stronger discriminators, they should be invested in more by strategists.

2.5   CONCLUSION

Current study attempted to assess factors, which discriminated between the intentions to use among females with regard to self-service banking technologies. In earlier stages of market development and product adoption, companies can segment the market on the basis of benefit bases like transaction costs and convenience. However, as the market concentration in term of repeat customer increases the marketers have to pay attention to nuances of independent variables like gender. Banking and use of technologies in India are at a stage where in order to increase adoption they have to think in terms of segmenting and targeting market on the basis of gender.

The current study uses discriminant analysis to determine which factors act as strong discriminators among females for intention to use self-service banking. The results of the study indicate that marketing strategists should concentrate on improving self- efficacy perception of the female customers to improve adoption of technology or self-service banking followed by decreasing the perception of risk while targeting perceived ease of use and perceived usefulness. Furthermore, for this sample two key areas of differences, perceived risk and capability, emerge. Female respondents exhibited a lesser confidence in their capability to use technology of self-service banking (technology discomfort) and also perceived it to be more risky. The female respondents seemed to be in stronger need for hand holding and assurance that the technology will operate reliably.

This can be done by investing in strategies like free demos, personal contact, and encouraging trial of self-service banking services by female customers. Furthermore, promoting self-service banking among this group of respondents communication strategies like explanatory videos which help reduce uncertainty and negative consequences of the self-service failure would be more helpful. The results indicate that because of reduction in the technology discomfort, the uneasiness of perceived self-efficacy can be countered by these simple strategies.

Therefore, the current study suggests that marketing practices for self-service banking need to be different for males and females. The gender differences in the study show that communication strategies in order to create more favorable attitude in males should concentrate more on perceived ease of use, while for females it should target both perceived ease of use and perceived usefulness.

The results of the study indicate that there are significant differences across genders with respect to drivers, which influence perceived usefulness and perceived ease of use of self-service technologies in banking. These differences need to be accepted, understood, and catered to by bankers to increase overall attitude of the customers towards self-service banking.

The research has a major limitation that impacts generalization of findings The research report findings of a study were undertaken in a developing country from four tier II cities. It might have limited implications if the results are generalized to all the population. These findings might not apply to tier I cities. Future research can address these limitations and also perform a cross-cultural study. Moderating effect of age and education also need to be studied in future studies.

KEYWORDS

- **self-service banking India**
- **females**
- **discriminant analysis**
- **intention to use**

REFERENCES

Baron, R. M.; Kenny, D. A. The Moderator–Mediator Variable Distinction in Social Psychological Research: Conceptual, Strategic, and Statistical Considerations. *J. Pers. Soc. Psychol.* **1986,** *51*(6), 1173.

Bauer, R. A. Consumer Behavior as Risk Taking. In *Dynamic Marketing for a Changing World;* Hancock, R. S., Ed., 1960, 389–398.

Bendapudi, N.; Leone, R. P. Psychological Implications of Customer Participation in Co-Production. *J. Mark.* **2003,** *67*(1), 14–28.

Byrnes, J. P.; Miller, D. C.; Schafer, W. D. Gender Differences in Risk Taking: A Meta-Analysis. *Psychol. Bull.* **1999,** *125*(3), 367.

Chau, P. Y. An Empirical Assessment of a Modified Technology Acceptance Model. *J. Manage. Inf. Syst.* **1996,** *13*(2), 185–204.

Chen, G.; Gully, S. M.; Eden, D. Validation of a New General Self-Efficacy Scale. *Organ. Res. Methods* **2001,** *4*(1), 62–83.

Chiu, Y.-B.; Lin, C.-P.; Tang, L.-L. Gender Differs: Assessing a Model of Online Purchase Intentions in E-Tail Service. *Int. J. Serv. Ind. Manage.* **2005,** *16*(5), 416–435.

Davis, F. D. *A Technology Acceptance Model for Empirically Testing New End-User Information Systems: Theory and Results.* Doctoral dissertation, Massachusetts Institute of Technology, 1985.

Davis, F. D. Perceived Usefulness, Perceived Ease of Use, and User Acceptance of Information Technology. *MIS Q.* **1989,** *13*(3), 319–340.

Davis, F. D. User Acceptance of Information Technology: System Characteristics, User Perceptions and Behavioral Impacts. *Int. J. Man-Mach. Stud.* **1993,** *38*(3), 475–487. DOI: http://dx.doi.org/10.1006/imms.1993.1022.

Durndell, A.; Haag, Z. Computer Self Efficacy, Computer Anxiety, Attitudes towards the Internet and Reported Experience with the Internet, by Gender, in an East European Sample. *Comput. Hum. Behav.* **2002,** *18*(5), 521–535.

Edgett, S.; Parkinson, S. Marketing for Service Industries-A Review. *Serv. Ind. J.* **1993,** *13*(3), 19–39.

Elliott, K. M.; Hall, M. C. Assessing Consumers' Propensity to Embrace Self-Service Technologies: Are There Gender Differences? *Mark. Manage. J.* **2005,** *15*(2), 98–107.

EY. Banking on Technology Perspectives on the Indian Banking Industry. January 2014. http://www.ey.com/Publication/vwLUAssets/EY-Banking-on-Technology/$FILE/ EY-Banking-on-Technology.pdf (accessed Jan 1, 2015).

Featherman, M. S.; Pavlou, P. A. Predicting E-Services Adoption: A Perceived Risk Facets Perspective. *Int. J. Hum.-Comput. Stud.* **2003,** *59*(4), 451–474.

Field, A. *Discovering Statistics using IBM SPSS Statistics*; Sage: 2013.

Fishbein, M. A Theory of Reasoned Action: Some Applications and Implications, 1979. *Nebraska Symposium on Motivation*, 27, 65–116.

Fishbein, M.; Ajzen, I. Belief, Attitude, Intention and Behavior: An Introduction to Theory and Research,1975. Addison-Wesley Pub (Sd), Addison-Wesly series in social psychology.

Garbarino, E.; Strahilevitz, M. Gender differences in the perceived risk of buying online and the effects of receiving a site recommendation. *J. Bus. Res.* **2004,** *57*(7), 768–775.

Gefen, D.; Straub, D. W. Gender Differences in the Perception and use of E-mail: An Extension to the Technology Acceptance Model. *MIS Q.* **1997,** *21*(4), 389–400.

Goodhue, D. L.; Thompson, R. L. Task-Technology Fit and Individual Performance. *MIS Q.* **1995,** *19*(2), 213–236.

Harris, C. R.; Jenkins, M.; Glaser, D. Gender Differences in Risk Assessment: Why do Women Take Fewer Risks than Men. *Judgment and Decision Making* **2006,** *1*(1), 48–63.

Judd, C. M.; Kenny, D. A. Process Analysis Estimating Mediation in Treatment Evaluations. *Eval. Review* **1981,** *5*(5), 602–619.

Karjaluoto, H.; Riquelme, H. E.; Rios, R. E. The Moderating Effect of Gender in the Adoption of Mobile Banking. *Int. J. Bank Mark.* **2010,** *28*(5), 328–341.

Kesharwani, A.; Bisht, S. S. The Impact of Trust and Perceived Risk on Internet Banking Adoption in India: An Extension of Technology Acceptance Model. *Int. J. Bank Mark.* **2012,** *30*(4), 303–322.

King, W. R.; He, J. A Meta-Analysis of the Technology Acceptance Model. *Inf. Manage.* **2006,** *43*(6), 740–755.

Lee, M.-C. Factors Influencing The Adoption of Internet Banking: An Integration of TAM and TPB with Perceived Risk and Perceived Benefit. *Electron. Commerce Res. Appl.* **2009,** *8*(3), 130–141.

Lee, Y. H.; Hsieh, Y. C.; Hsu, C. N. Adding Innovation Diffusion Theory to the Technology Acceptance Model: Supporting Employees' Intentions to Use E-Learning Systems. *J. Educ. Technol. Soc..* **2011,** *14*(4):124.

Legris, P.; Ingham, J.; Collerette, P. Why do People Use Information Technology? A Critical Review of The Technology Acceptance Model. *Inf. Manage.* **2003,** *40*(3), 191–204.

Meuter, M. L.; Ostrom, A. L.; Roundtree, R. I.; Bitner, M. J. Self-Service Technologies: Understanding Customer Satisfaction with Technology-Based Service Encounters. *J. Mark.* **2000,** *64*(3), 50–64.

Meuter, M. L.; Bitner, M. J.; Ostrom, A. L.; Brown, S. W. Choosing among Alternative Service Delivery Modes: An Investigation of Customer Trial of Self-Service Technologies. *J. Mark.* **2005,** *69*(2), 61–83.

Mun, Y. Y.; Hwang, Y. Predicting The Use of Web-Based Information Systems: Self-Efficacy, Enjoyment, Learning Goal Orientation, and The Technology Acceptance Model. *Int. J. Hum.-Comput. Stud.* **2003,** *59*(4), 431–449.

Nunnally, J. C. *Psychometric Theory,* 2nd Ed.; McGraw-Hill: New York, 1978.

Ong, C.-S.; Lai, J.-Y. Gender Differences in Perceptions and Relationships Among Dominants of E-Learning Acceptance. *Comput. Hum. Behav.* **2006,** *22*(5), 816–829.

Parasuraman, A. Technology Readiness Index (TRI) A Multiple-Item Scale to Measure Readiness to Embrace New Technologies. *J. Serv. Res.* **2000,** *2*(4), 307–320.

Rogers, E. M. *Diffusion of Innovations;* Simon and Schuster: New York, 2010.

Schwarzer, R.; Bassler, J.; Kwiatek, P.; Schroder, K.; Zhang, J. X. The Assessment of Optimistic Self-Beliefs: Comparison of The German, Spanish, and Chinese Versions of The General Self-Efficacy Scale. *Appl. Psychol.* **1997,** *46*(1), 69–88.

Sharma, R. R.; Chander, S. Consumer Psychographics and Surrogate Advertising: An Application of Multiple Discriminant Analysis. *Icfai J. Consum. Behav.* **2007,** *II*(5), 25–47.

Taylor, S.; Todd, P. A. Understanding Information Technology Usage: A Test of Competing Models. *Inf. Syst. Res.* **1995,** *6*(2), 144–176.

Venkatesh, V. Determinants of Perceived Ease of Use: Integrating Control, Intrinsic Motivation, and Emotion into the Technology Acceptance Model. *Inf. Syst. Res.* **2000,** *11*(4), 342–365.

Venkatesh, V.; Davis, F. D. A. Model of the Antecedents of Perceived Ease of Use: Development and Test*. *Decis. Sci.* **1996,** *27*(3), 451–481.

Venkatesh, V.; Morris, M. G.; Ackerman, P. L. A. Longitudinal Field Investigation of Gender Differences in Individual Technology Adoption Decision-Making Processes. *Organ. Behav. Hum. Decis. Processes* **2000,** *83*(1), 33–60.

Vijayasarathy, L. R. Predicting Consumer Intentions to Use On-Line Shopping: The Case for an Augmented Technology Acceptance Model. *Inf. Manage.* **2004,** *41*(6), 747–762. DOI: http://dx.doi.org/10.1016/j.im.2003.08.011.

Wang, Y. S.; Wu, M. C.; Wang, H. Y. Investigating the Determinants and Age and Gender Differences in The Acceptance of Mobile Learning. *Br. J. Educ. Technol.* **2009,** *40*(1), 92–118.

Yiu, C. S.; Grant, K.; Edgar, D. Factors Affecting the Adoption of Internet Banking in Hong Kong—Implications for the Banking Sector. *Int. J. Inf. Manage.* **2007,** *27*(5), 336–351.

Zeithaml, V. A.; Parasuraman, A.; Berry, L. L. Problems and Strategies in Services Marketing. *J. Mark.* **1985,** *49*(2), 33–46.

Zmud, R. W. Individual Differences and MIS Success: A Review of the Empirical Literature. *Manage. Sci.* **1979,** *25*(10), 966–979.

CHAPTER 3

ANTI-SEXUAL HARASSMENT POLICIES AND HR PRACTICES: PREVENTION AND PREPARATION

RUCHI JOSHI[1,*]and MADHU JAIN[2]

[1]Counseling Psychologist, Udaipur, India,
*E-mail: ruchijoshi43@yahoo.com

[2]Department of Psychology, University of Rajasthan, Jaipur, India,
E-mail: madhujain28@gmail.com

CONTENTS

ABSTRACT

The present chapter is an attempt to address various issues related to sexual harassment prevalent worldwide specifically against women, the efforts and the long battle ensued by activists from all over the world in drawing attention toward the severity and consequences of sexual harassment at individual, organization, and economic levels globally. It also intends to shed light on various forms of sexual harassment faced by females at workplace worldwide as this form of discrimination is found to be in existence not only in underdeveloped economies and third world countries, but the menace is very much present in many of the developed economies and in so-called superpowers as well. The psychology of the perpetuators and that of victims will be dwelled upon along with various policies and ideologies related to sexual harassment in various parts of the world as what all behavior will constitute as "sexual harassment," the legal safeguards for victims and repercussions against perpetuators. The role of human resources (HR) department universally in combating sexual harassment and creating a safe discrimination-free environment will be discussed at length as its scope is all-inclusive and wide reaching for it shoulders the responsibility to manage the most valuable assets for any organization, that is, the HR to ensure overall development of the organization by discussing a few approaches and making some suggestion for creation of a discrimination-free, gender sensitive, productive and absolutely professional work environment to make the world free of menace of sexual harassment.

3.1   INTRODUCTION

The world could not get enough scoop of the famous or rather *infamous* trial that went against Bill Clinton sometime during the beginning new millennium when after scores of women including Monika Lewinsky, Jaunita Broaddrick and Paula Jones, etc. who alleged that the former American President had more than professional interest in them and had suffered sexual misconduct at his hands. In the process of allegations, denials, counter allegations, entire world was stunned to witness sexual harassment trial against one of the most powerful officeholders, which proved that the menace of sexual harassment had its presence around all nooks and corners of the world.

Recently sexual assault allegations against media mogul Harvey Weinstein exploded into public view. Since then there have been new allegations made against other powerful men in various industries seemingly almost every day. (Retrieved from edition.cnn.com/2017/10/25/us/list-of-accused-afterwienstein-scandal-trnd-/index.html, on 20th November 2017). These revelations not only made organizations to look at their anti-sexual policies again but also led to a worldwide online campaign #MeToo where women all over the world were asked to share stories of any form of sexual harassment they faced at any point of time. In a disgustingly shameful series in which women all over relived the bitter, torturous, and humiliating moments of abuse by their perpetuators.

Indian media was abuzz with charges of sexual harassment levied by a research associate against the Chairman of Intergovernmental Panel for Climate Change (IPCC) Rajendra Kumar Pachauri in 2015. Pachauri is not only a highly reputed academician but also a joint noble laureate along with U.S. Vice President Al Gore.

England's Detective Chief Inspector (DCI) Wayne Jones, who has worked on some of the country's biggest murder cases, was suspended after a married junior officer reported him to West Midlands Police bosses in January 2014. The victim complained her life was made "hell" with persistent sexually inappropriate comments after she rejected his advances between 2012 and 2013. Her actions prompted three other female colleagues to come forward who also claimed DCI Jones had made repeated unwanted advances and lewd comments.

A leading French newspaper reported that 17 women who have served as ministers in France say they will no longer be silent about sexual harassment in politics. (www.bbc.com)

Apart from these famous and well-known cases related to sexual harassment, there are innumerous cases of harassment at work in right from multinational companies to small-scale industries to unskilled and unaccounted labor which goes unnoticed along with its ugly repercussions on physical, emotional, psychological, and occupational state of bearer. Despite hue and cry from half the population worldwide, creation of laws and provisions and much awareness related to sexual harassment, there is still a stratum which believes that workplace sexual harassment either does not exist or is an understatement.

There has been a disagreement about whether, and which, sexual behaviors are harassing and why. Many people are skeptical—if not resentful—toward the idea that sexual harassment is a serious issue and form of sex discrimination (cf. Berdahl et al., 1996) Some have argued that sexual banter and jokes provide a fun and jovial atmosphere at work and that sexual flirtation and invitations can be flattering and result in love and romance. It has also been suggested that "sanitizing" the workplace of sexual behavior hands companies an age-old excuse to keep the sexes separate and unequal at work. In their study Berdahl and Aquino (2009) found that exposure to sexual behavior at work predicted negative employee work and psychological well-being, even for employees who said they enjoyed the experience.

Due to activism exhibited by females and like-minded males, in the past several years, sexual harassment has received increasing attention from researchers and organizations alike once virtually unknown in the scientific literature, the topic currently yields over 500 references, the great majority published in the last 5 years; incidence and prevalence studies abound, and edited volumes and special issues of journals have begun to appear (Borgida and Fiske, 1995; O'Donohue, 1997; Pryor and McKinney, 1995; Stockdale, 1996; Tinsley and Stockdale, 1993). Considerable data have accumulated confirming prevalence of sexual harassment worldwide, especially against women. Although "sexual harassment" in itself is not gender specific and both the genders are equally vulnerable in this regard, in the present chapter, we will take into account harassment related to women only, because while sexual harassment may be and is experienced by both genders, it is however the "fairer" sex who is at receiving end of harassment and its unfortunate consequences most of the times.

Women worldwide have stormed out of safety sanctuaries and carved a niche for themselves in fields hitherto was a foreign territories for them. There is absolutely no field which has not felt strong presence of the half *population* striving hard to improve its lot. With advances in technology, healthcare, education, accessibility to remote areas and innumerable efforts have been made by many visionaries to enhance position of women and more than that strong aspiration among women to abandon every shackle; physical, mental or emotional to make long strides in path of development.

Having proved their mettle worldwide, having splintered the formidable glass ceiling when women thought they have finally arrived, a rude shock awaited them in the form of sexual harassment in various forms. All over the world, organizations have sought to uproot the evil, at least in principle. There is a lot left to be done in practice.

Sexual harassment is a violation of an individual's human rights and a prohibited form of violence against women and in some cases against men in many countries. However, sexually harassing conduct causes much devastating physical and psychological injuries to a large percentage of women in workplaces around the world. Harassment directed against women in the workplace by their supervisors, fellow employees, or third parties interferes with the integration of women in the workforce, reinforces the subordination of women to men in society, violates women's dignity and creates a health and safety hazard at work.

Women's advocates around the world work to further women's right to be free from sexual harassment. Critical to these efforts to combat sexual harassment has been the growing recognition of sexual harassment as a form of violence against women which violates women's human rights. States are obligated under international law to take effective steps to protect women from violence and to hold harassers and/or their employers accountable for sexual harassment in the workplace.

The term "sexual harassment" first came into use in the late 1970s in the United States. The term's origins are generally traced to a course on women and work taught by Lin Farley at Cornell University. In 1979, Catherine MacKinnon, a legal scholar from the United States, made the first argument that sexual harassment is a form of sex discrimination prohibited by the constitution and civil rights laws of the United States.

MacKinnon argued for the legal recognition of sexual harassment as sex discrimination in her 1979 book Sexual Harassment of Working Women. In the book, MacKinnon states that because of the traditional gender roles of our society, women disproportionately occupy inferior positions in the workplace. One psychologist writing on the subject concurred with MacKinnon, seeing sexual harassment, "as a form of sex discrimination that keeps the sexes separate and unequal at work" (Berdahl, 2007).

MacKinnon (1979) argues that "intimate violations" of women by men were "sufficiently pervasive" as to make the practice nearly invisible (p. 1). She also states that internalized power structures within the workplace kept anyone from discussing sexual harassment, making it "inaudible." In her words, the abuse was both acceptable for men to perpetuate and a taboo that women could not confront either publicly or privately. MacKinnon states that the "social failure" to address these pervasive intimate violations hurt women in terms of the economic status, opportunity, mental health, and self-esteem. Many believe that sexual harassment is about the abuse of power, others believe it is about access to sexual favors, and still others believe that sexual harassment is about access to power and sex. In legal terms, sexual harassment is divided into two main categories that are as follows.

3.1.1   QUID PRO QUO

Quid pro quo harassment occurs when an employee is made to submit to some form of sexual advance in order to obtain a benefit (e.g., a promotion) or to avoid a burden (e.g., being fired). In such cases, sexual harassment is considered sex discrimination because presumably the demand would not have been made if the employee were of the opposite sex (Browne, 2006). Initially, researchers and courts believed that this type of harassment was motivated by sexual desire, but research has subsequently suggested that it is instead meant to assert dominance over or derogate the target (Berdahl, 2007).

3.1.2   HOSTILE ENVIRONMENT

Hostile environment harassment occurs when the work environment is "permeated with sexuality" or "discriminatory intimidation, ridicule, and insult" (Smith et al., 2011). Within this type of harassment, the victim does not claim specific harassment, but rather that the general work environment is discriminatory. Generally, it is believed that this type of harassment seeks to undermine and humiliate its target and is likely to be motivated by sexual hostility rather than sexual desire (Berdahl, 2007).

The concept of sexual harassment, in its modern understanding, is a relatively new one, but as mentioned above, is dating from the 1970s onwards; although other related concepts have existed prior to this in many cultures. The term "sexual harassment" was used in 1973 in "Saturn's Rings," a report authored by Mary Rowe to the then President and Chancellor of MIT about various forms of gender issues. Rowe has stated that she believes and was not the first to use the term, since sexual harassment was being discussed in women's groups in Massachusetts in the early 1970s, but that MIT may have been the first or one of the first large organizations to discuss the topic (in the MIT Academic Council), and to develop relevant policies and procedures. MIT at the time also recognized the injuries caused by racial harassment and the harassment of women of color, which may be both racial and sexual. The President of MIT also stated that harassment (and favoritism) are antithetical to the mission of a university as well as intolerable for individuals (Wikipedia).

Since then many international bodies, national legislatures and courts have prohibited sexual harassment but have not agreed on a universal definition of the term.

In India, The *Sexual Harassment of Women at Workplace (Prevention, Prohibition, and Redressal) Act, 2013* seeks to protect women from sexual harassment at their place of work. It was passed on February 26, 2013 and came into force from December 9, 2013. This statute superseded the Vishakha Guidelines for prevention of sexual harassment introduced by the Supreme Court of India. The Act ensures that women are protected against sexual harassment at all the work places, be it in public or private. This will contribute to the realization of their right to gender equality, life, and liberty and equality in working conditions everywhere. The sense of security at the workplace will improve women's participation in work, resulting in their economic empowerment and inclusive growth.

The act makes it obligatory for all organizations, public or private to establish an anti-sexual harassment committee to which complains related to sexual harassment will be referred and proceeding are to take place are according to the law of the land and decision must be delivered within 90 days.

Since then many international bodies, national legislatures and courts have prohibited sexual harassment but have not agreed on a universal definition of the term. Noncompliance with the provisions of the Act shall be punishable with a fine of up to 50,000. Repeated violations may lead to higher penalties and cancellation of license or registration to conduct business. Government can order an officer to inspect workplace and records related to sexual harassment in any organization.

The United Nations General Recommendation 19 to the Convention on the Elimination of all Forms of Discrimination Against Women defines sexual harassment of women to include:

> "Such unwelcome sexually determined behavior as physical contact and advances, sexually colored remarks, showing pornography and sexual demands, whether by words or actions. Such conduct can be humiliating and may constitute a health and safety problem; it is discriminatory when the woman has reasonable ground to believe that her objection would disadvantage her in connection with her employment, including recruitment or promotion, or when it creates a hostile working environment."

In the European Union, there is a directive on sexual harassment. The "Directive 2002/73/EC—equal treatment of 23 September 2002 amending Council Directive 76/207/EEC on the implementation of the principle of equal treatment for men and women as regards access to employment, vocational training and promotion, and working conditions" states.

For the purposes of this Directive, the following definitions shall apply: "sexual harassment: where any form of unwanted verbal, non-verbal or physical conduct of a sexual nature occurs, with the purpose or effect of violating the dignity of a person, in particular when creating an intimidating, hostile, degrading, humiliating or offensive environment" harassment and sexual harassment within the meaning of this Directive shall be deemed to be discrimination on the grounds of sex and therefore prohibited.

There are a few common elements in definitions of sexual harassment worldwide as mentioned above. Generally speaking, behavior constituting sexual harassment in the workplace must:

1. Occur in the place of work or in a work-related environment
2. Occur because of the person's sex and or it is related to or about sex
3. Be unwelcome, unwanted, uninvited, not returned, not mutual
4. Affect the terms or conditions of employment (*quid pro quo* sexual harassment) or the work environment itself (hostile work environment sexual harassment).
5. Include peering, ogling, passing remarks, touching, making obscene sounds brushing against body parts, forceful demands of dating or meeting outside work and so forth.

3.2   REVIEW OF LITERATURE

Having discussed what constitutes sexual harassment, it would only fall in place if it is explored why any harassment sexual or otherwise takes place at all? Quoting the work of Martha Chamallas (2003), one of the theories that emerge is that of subordination of women to which was first introduced in United States in 1970s. This idea associated sexual harassment against women, perpetuation of gender stereotypes, and assertion of economic power over women are all the indicators of phenomena which support and supplicate the idea that men are above women and therefore entitled to have their way in case of female exploitation.

Extensive research on such harassment suggests that it has less to do with sex than with power. It is a way to keep women in their place; through harassment men devalue a woman's role in the work place by calling attention to her sexuality. "Sexual harassment is a subtle rape, and rape is more about fear than sex," said Dr. John Gottman, a psychologist at the University of Washington. "Harassment is a way for a man to make a woman vulnerable." While sexual harassment may at first glance be taken as simple social ineptness or as an awkward expression of romantic attraction, researchers say that this view is wrong and pernicious because it can lead the women who suffer harassment to blame themselves, believing that something in their dress or behavior might have brought the unwanted attention.

In fact, only about 25% of cases of sexual harassment are botched seductions, in which the man "is trying to get someone into bed," said Dr. Louise Fitzgerald, a psychologist at the University of Illinois. "And in less than 5% of cases the harassment involves a bribe or threat for sex, where the man is saying, 'If you do this for me, I'll help you at work, and if you don't, I'll make things difficult for you.'" "The rest," she said, "are assertions of power." All the signs of a tactic.

Researchers say, the use of harassment as a tactic to control or frighten women, explains why sexual harassment is most frequent in occupations and work places where women are new and are in the minority. In fact, no matter how many men they encounter in the course of their work, women who hold jobs traditionally held by men are far more likely to be harassed than women who do "women's work."

For example, a 1989 study of 100 women working in a factory found that those who were machinists, not a job traditionally held by women, reported being harassed far more than those on the assembly line, where more women work. Women in both groups encountered about the same number of men at work.

"On all 28 items of a sexual harassment scale, ranging from lewd remarks to sexual assault, the women machinists had the highest scores," said Dr. Nancy Baker and associates (1987), a psychologist in Los Angeles who conducted the study. "Among women in white-collar jobs, the same holds true. The more nontraditional the job for women, the more is the sexual harassment. Women surgeons and investment bankers rank among the highest for harassment."

The style of harassment also is likely to differ among professionals and blue-collar workers. "In the blue-collar work place there's often a real hostility towards women," said Dr. Fitzgerald. "Men see women as invading a masculine environment." These are guys whose sexual harassment has nothing whatever to do with sex. They're trying to scare women off a male preserve (Goleman 1991).

In cultures where it is, or was until recently, "OK" to discriminate against people because they are different (in terms of gender, race, culture, religion, lifestyle, political conviction or whatever), the abuse of power or humiliation that is typical of sexual harassment will not be unusual. Harassment is often closely linked to prejudice in general, and to sexist attitudes.

Men who were brought up with macho beliefs such as "real men pinch bottoms," "girls were made to hug and kiss," "the more, the merrier," easily carry these social values into the workplace, and treat their female colleagues accordingly in case of female sexual harassment. Such men often are of the opinion that women take their harassment as a compliment.

Economic causes also account for silence of women, consequently encouragement to sexual harassment direct or indirect. Several studies have found that only 3% of women who have been sexually harassed make a formal complaint. "We find that close to 90% of women who have been sexually harassed want to leave, but can't because they need their job" (Goleman 1991).

One of the leading causes of sexual harassment prevalent at work place is due to the absence of a clear, well-defined policy on sexual harassment at work place. In the absence of one or in case of policy being ambiguously framed, there are chances of sexual harassment being practiced directly or indirectly against any gender.

Since Sexual harassment has serious consequences for both victims who face it first hand or for other who see perpetuator walking harmlessly in premises. Studies quote that sexual harassment whether direct or indirect may lead to absenteeism, lower levels of efficiency, physiological and psychological symptoms like nausea, dizziness, profound sweating, and high or low blood pressure, and so forth. Psychological symptoms include inability to concentrate, lack of self-confidence, distress, and so forth. Mental disorders like depression, Post-traumatic stress disorder and personality disorders are also consequences of sexual harassment.

Pryor (2010) studied the prevalence, dimensions, and correlates of psychological harm that women experience as a result of sexual harassment in the workplace. Data were collected from a worldwide survey of sexual harassment in the active duty U.S. military. The scientifically selected sample included over 10,000 working military women. Four general types of negative psychosocial reactions were identified among victims of sexual harassment: productivity problems, attitudes toward the organization, emotional reactions, and relations with family. Analyses explored the relations of these psychosocial reactions to (a) characteristics of the harassing behavior (what happened and who did it), (b) characteristics of the victim, (c) characteristics of the organizational climate in which the harassment took place, and (d) the victim's coping responses.

Owing to the losses incurred by any organization directly or indirectly, economically, in terms of lowered quality of work, adversely affected morale of employees, bearing legal costs and finally attrition rates, there is a serious need of clear well-defined policies related to sexual harassment and also the for the HR department (HRD) to proactively, in association with the higher-ups in creating and implementing clear, well-defined anti-sexual harassment policies, consequences if found guilty of either committing or false accusations the act and thereafter not only make employees aware of the policies, implement it but also ensure that proper follow-up and counseling services are being provided to those affected by it.

Miner-Rubino and Cortina (2007) examined two indicators of a hostile interpersonal workplace climate for women-observed hostility (i.e., incivility and sexual harassment) toward women and perceived organizational unresponsiveness to sexual harassment and how they relate to the well-being and withdrawal for employees. Participants included 871 female and 831 male employees from a public university. According to structural equation analyses, observing hostility toward women and perceiving the organization as lax about harassment predicts lower well-being, which translates into higher organizational withdrawal for both female and male employees. Results hold even after controlling for personal mistreatment, negative affectivity, and observed hostility toward men. These findings suggest that working in a misogynistic environment can have negative effects for all employees.

Woodzicka and LaFrance (2005) conducted a study to examine real-time consequences of subtle sexual harassment in a job interview using objective indicators of job performance. Fifty women were recruited for a job interview. Participants were randomly assigned to one of two interview conditions during which they were asked either three sexual or nonsexual questions interspersed with standard interview questions. In the former, women applicants spoke less fluently, gave lower quality answers, and asked fewer job relevant questions than did those in the nonsexual interview. It thus appears that even relatively mild harassment disrupts immediate performance.

3.3 SUGGESTIONS

The question now arises as to how the HRD can assist in creating a "safe and anti-discriminatory" work environment? How can the HRD ensure whether a genuine case of sexual harassment is been registered or is there an attempt to make someone a scapegoat for settling personal scores or is used as means for achieving some reprobate ends? If some innocuous acts have been misinterpreted by someone? And last but not the least if there has been an actual case of sexual harassment taking place or an attempt has been made for the same what will be the preventive measures to be taken by the HRD? Described below are some means and approaches to combat the menace of sexual harassment from raising its ugly head, which can be employed by the HRD in any organization.

Proactive role in prevention of sexual harassment is the first and most obvious task of HRD in preventing sexual harassment at work by establishing an anti-sexual harassment committee constituting of heads of the organizations/departments/legal experts and counselors (internal or external) with an appropriate number of males and females in it. Next the committee along with its policies and agenda should be brought to everyone's knowledge. This can be attained through group meetings, presentations, seminars and group discussions. New recruits should be briefed about the organization's dogged commitment towards an anti-sexual harassment work environment as well as knowledge about presence of a formal and acting committee for the purpose should be provided. Most of the time employees are not aware as to which behavior constitutes sexual harassment. In a study Joshi and Jain (2016) conducted on front-office executives in Jaipur, it was found that almost all the participants were under impression that only penetrative sex constitutes sexual harassment.

3.3.1 ASSISTANCE WITH SEXUAL HARASSMENT CONCERNS

The committee along with HRD should be responsible for ensuring and monitoring the Organization's compliance with state nondiscrimination laws. However, a discrimination-free environment is the responsibility of every member of the community. The HRD must encourage persons who believe they have experienced or witnessed sexual harassment to come forward promptly with their inquiries, reports, or complaints and to seek assistance without the fear of being judged, ridiculed or threat to their employment.

3.3.2 CONFIDENTIAL COUNSELING

Information about or assistance with sexual harassment issues should be obtained from a variety of sources. Prior to or concurrent with lodging a sexual harassment complaint, individuals may find it helpful to consult with a committee member or a counselor along with HRD to seek assistance. All information shared should remain confidential to the extent permitted by law and the organization's policy. Discussions with these representatives should not be considered a report to the HRD regarding the problematic behavior and should not, without additional action by the complainant, result in intervention or corrective action.

3.3.3 LODGING A COMPLAINT

An individual should be able to complain to the organization about alleged sexually harassing behavior to the committee or to office HRD. Provisions of information dissention about reporting such cases and about the standard procedure that follows should be made through trainings and meeting.

3.3.4 INVESTIGATION AND INVESTIGATION PROCEDURES

The HRD should handle sexual harassment complaints consistently with procedural guidelines developed to ensure prompt and equitable resolution of such complaints. Complainants and respondents should be given copies of the procedural guidelines, and the guidelines should also be made readily available to the organization. The matter should then proceed to investigation or other form of effective and fair review. The purpose of an investigation should include interviewing the parties and witnesses, to gather and assess evidence (inputs from Michigan University).

Human resource department must establish a specific procedure for sexual harassment complaints or, alternatively, use the procedure that is already in place for other types of work-related grievances. However, sexual harassment complaints are often complex, sensitive and potentially volatile. Anyone who has responsibility for dealing with them will require specialist expertise and should receive appropriate training. HRD should ensure that their organization's complaint procedures:

1. Are clearly documented
2. Are explained to all employees
3. Offer both informal and formal options
4. Address complaints in a manner which is fair, timely and confidential
5. Are based on the principles of natural justice
6. Are administered by trained personnel
7. Provide clear guidance on internal investigation procedures and record keeping
8. Advise a complainant that they can pursue the matter externally with a State or Territory anti-discrimination body or, if it appears to be a criminal matter, the police
9. Give an undertaking that no employee will be victimized or disadvantaged for making a complaint
10. Are regularly reviewed for effectiveness
11. Should make preemployment screening and reference checks regarding problems with interpersonal behavior and relationships

The department must also ensure that if found guilty, the perpetuator of the behavior must strictly be dealt with according to the policies of the organization and the law of state to ensure the ideal of safe, discrimination-free environment is intact and also to inspire confidence among work force, rather than sugar coating the matter or pushing it under the carpet.

Another most important responsibility the HRD must undertake is that of helping in creation of not only a gender-neutral environment, but also of a friendly environment for both genders where a sense of confidence and comradeship is felt rather than lack of trust and suspicion, The predominant apprehension that most men had was the fact that the powers provided by the law can be used lethally by ill-willed women to destroy careers and hard-earned reputation.

A lot of men are scared of what can happen if women resort to abusing this power and using the law in malicious ways. Given that there are several instances of misuse of women-friendly provisions in criminal law, especially in cases of souring marital relationships, men fear that the same applies to anti-sexual harassment laws.

However, the truth is the changing mindset of law enforcement authorities and legal knowledge can minimize or eliminate misuse of anti-sexual harassment laws completely.

The law aims to equip organizations to handle a workplace-related evil, not give women undue power or to put men in fear. It is true that women-friendly laws have been misused in the past, but courts and police have become sensitive to this and they are more careful now to check the genuineness of complaints before about taking adverse action.

It is not that a man will immediately be punished once a complaint is filed against him. Organizations are aware of the possibility of misuse and don't want to encourage it.

Several male employees said that they had contemplated avoiding collaborations with women on projects and so forth in fear of the anti-sexual harassment law being invoked against them falsely. However, it must be pointed out that building a conducive workplace environment is in the interest of everyone, both men and women. The law cannot work as effectively without male support. If we allow sexual harassment to exist at the workplace and not take steps against it, we are paving the way for other kinds of harassment and discrimination to also creep in.

There should be provisions for trainings; legal and psychological by the HRD for all the employees right from top-notch to the lowest related to sexual harassment.

There should also be provisions of post-harassment psychological counseling at company's expenses in case there has been a case of sexual harassment to help the victim rebuild her confidence, self-esteem and professional efficiency.

Human resources are the most valuable assets for any organization and HRD constitutes the select, trained and able individuals who ensure that these assets must work to the best of their ability and strive their full potential, thereby contributing to the growth of the organization and their personal growth. The department is entrusted with creating such favorable conditions for the employees to attain the same. Creation of a safe, discrimination-free, anti-sexual, and harassment-free work environment is one of the requisites for that. The HRD worldwide should strive to create a safe, healthy and conducive environment for all employees irrespective of culture, race, nationality, preferences and most importantly gender.

3.4   CONCLUSION

Sexual harassment has a severely negative impact on working population worldwide and its direct and indirect consequences result in absenteeism, frequent medical leaves, lack of concentration, lower outputs, high attrition rates and higher costs to organization. Personally, it can lead to distress, anxiety, and physical symptoms such as high or low blood pressure, lack of appetite, giddiness, dizziness and so forth. Although it is agreed upon that sexual harassment is not gender specific and its roots lie deep beyond just "sexual" aspects and it actually is related to power dynamics, where higher-ups take advantage of their positions to exploit those subordinate to them. However, it is usually the "fairer sex" who is generally at the receiving ends of sexual harassment directly or indirectly. The responses of victims are influenced by the amount of support and understanding they receive from significant others and employers. The extent of emotional, physical, and psychological damage depends on the responsiveness of other people and the organization for which the woman works. However, the effect on the morale of all employees can also be serious and not just exclusively on the victim. Both men and women in a workplace can find their work disrupted by sexual harassment even if they are not directly involved. Sexual harassment can have a demoralizing effect on everyone within range of it, and it often negatively impacts company productivity on the whole. HRD can play a pivotal role in establishing a fair, productive safe and discrimination-free environment by taking into account several measures like creation of an anti-sexual harassment committee, formulating a clear, specific and documented policy against any sort of discrimination training sessions and activities related to information dissention etc. HRD should also entrust itself with the responsibility of carrying out an affair and unbiased investigation and if found guilty, appropriate action should be taken against the culprit in accordance with the law of the land. The victim should be provided counseling services to ensure psychological and emotional well-being as incidences of sexual harassment are not conducive only to physical, emotional health but also to performance in an organization.

KEYWORDS

- **sexual harassment**
- **work environment**
- **worldwide scenario**
- **HR practices**
- **psychological trauma**
- **assistance**
- **procedures**
- **counseling**

REFERENCES

Berdahl, J. L.; Magley, V.; Waldo, C. R. The Sexual Harassment of Men? Exploring the Concept with Theory And Data. *Psychol. Women Q.* **1996,** *20,* 527–547.

Berdahl, J. L. Harassment Based on Sex: Protecting Social Status in the Context of Gender Hierarchy. *Acad. Manage. Rev.* **2007,** *22,* 641–658.

Berdahl, J.L.; Aquino, K. Sexual Behaviour at Work: Fun or Folly. *J. Appl. Soc. Psychol.* **2009,** *94,* 34–47.

Borgida, E.; Fiske, S. T. By Way of Introduction. *J. Soc. Issues* **1995,** *51,* 1–10.

Browne, K. R. Sex, Power, and Dominance: The Evolutionary Psychology of Sexual Harassment. *Managerial Decis. Econ.* **2006,** *27,* 145–158.

Chamallas, M. *Introduction to Feminist Legal Theory.* Aspen Publishers: New York, 2003.

Fitzgerald, L. F.; Dragow, F.; Hulin, C. L.; Gelfand, J. J.; Magley, V. J. Antecedents and Consequences of Sexual Harassment in Organizations: A Test of an Integrated Model. *J. Appl. Psychol.* **1997,** *82*(4), 578–589

French Female Ministers Decry Sexual Harassment, 2016. http://www.bbc.com/news/world-europe-36297081 (accessed Sept 4, 2016).

Goleman, D. *Sexual Harassment: It's About Power, Not Lust.* 1991, http://www.nytimes.com/1991/10/22/science/sexual-harassment-it-s-about-power-not-lust.htmlon (accessed Sept 4, 2016).

Joshi. R.; Jain. M. Knowledge, Attitude and Practices Related to Workplace Harassment Among Women: Empowerment Through Knowledge Dissemination. In *Human Rights and Disadvantaged Groups: Status and Challenges* Thanvi, S., Ed.; Suraj Sansthan: Jaipur (Raj.), 2016; pp 59–62.

MacKinnon, C. *Sexual Harassment of Working Women: A Case of Sex Discrimination;* Yale University Press: New Haven, *1979*.

Miner-Rubino, K.; Cortina, L. M. Beyond Targets: Consequences of Vicarious Exposure to Misogyny at Work. *J. appl. psychol.* **2007,** *92*(5), 1254–1269

O'Donohue, W. *Sexual Harassment: Theory, Research and Treatment;* Allyn & Bacon: Boston, 1997.

Pryor, J. B. The Psychological Impact of Sexual Harassment on Women in the U.S. Military. *Basic Appl. Psychol.* **2010,** *17,* 581–603.

Pryor, J. B.; McKinney, K. Research on Sexual Harassment: Lingering Issues and Future Directions. *Basic Appl. Soc. Psychol.* **1995,** *17,* 605–611

Smith, B. N.; Shipherd, J. C.; Schuster, J. L.; Vogt, D. S.; King, L. A.; King, D. W. Posttraumatic Stress Symptomatology As A Mediator of The Association Between Military Sexual Trauma And Post-Deployment Physical Health In Women. J. Trauma Dissociation, 4 sep 2016, *12*, 275–289. Retrieved from http://dx.doi.org/10.1080/15299732.201 1.55150.

Stockdale, M. S. *Sexual Harassment in the Workplace: Perspectives, Frontiers, and Response Strategies;* Thousand Oaks, CA: Sage, 1996.

Terpstra, D. E.; Baker, D. D. A Hierarchy of Sexual Harassment. *J. Psychol.* **1987,** *121,* 599–605.

Tinsley, H. E. A.; Stockdale, M. S. Sexual Harassment in the Workplace. *J. Vocational Behav.* **1993,** *42,* 1–4.

Webb, S. L. *Shockwaves: The Global Impact of Sexual Harassment*; MasterMedia Limited: New York, 1994.

Woodzicka, J. A.; LaFrance, M. The Effects of Subtle Sexual Harassment on Women's Performance in a Job Interview. *Sex Roles* **2005,** *53,* 67–77

CHAPTER 4

JOB SATISFACTION IN IT-ENABLED WORKING ENVIRONMENTS: COMPARATIVE ANALYSIS BETWEEN INDUSTRIES

ŽIVA VEINGERL ČIČ[1,*], SAMO BOBEK[2], and
SIMONA ŠAROTAR ŽIŽEK[3]

[1]Faculty of Economics and Business, University of Maribor, Razlagova, Maribor, Slovenia, *E-mail: zivana.veingerl1@um.si

[2]Faculty of Economics and Business, University of Maribor, Razlagova, Maribor, Slovenia, E-mail: samo.bobek@um.si

[3]Faculty of Economics and Business, University of Maribor, Razlagova, Maribor, Slovenia, E-mail: simona.sarotar-zizek@um.si

CONTENTS

ABSTRACT

PURPOSE

There is considerable debate among academics and business practitioners about key factors of existence and development of organizations and employees; of which the important one is the value of the use of social networking by organizational members. An important factor of the employee development is certainly employee satisfaction at the workplace, because it defines the way the employees experience the work, and because it influences people's attitude towards their jobs. Since working in the field of information technology (IT) and high technology is challenging, and the development cycle is changing rapidly, the satisfaction of employees at the workplace is very important for company success. Job satisfaction is also an important factor that shapes the future career of the individual. In this chapter, the characteristics of employee satisfaction are shown between the four countries in the field of IT covered by the survey.

METHODOLOGY/APPROACH

The exploratory study is based on a four-country online survey, including 1207 employees who fulfilled the two criteria; first one was having a job of at least 10 working hours per week and the second was having colleagues at work. Among the respondents 294 were from Austria, 283 from Germany, 369 from Slovenia, and 261 from Spain.

FINDINGS

The results show that the differences in employee's job satisfaction between IT and other sectors in four countries are not statistically significant.

ORIGINALITY/VALUE

This is the first study to analyze job satisfaction in four countries regarding sector/activity.

4.1 INTRODUCTION

Hi-tech professionals are increasingly important for the companies due to rapid development of the information technology (IT), to maintain or create their competitive advantage in the market. Many authors are confident that job satisfaction is the main driver of turnover among hi-tech professionals (Niederman and Sumner, 2004; Poulin, 1994; West and Bogumil, 2001).

Most studies have indicated that job satisfaction significantly influences organizational behavior and positively affects job performance and/or organizational commitment (Christen, et al., 2006; Cohrs et al., 2006; Rayton, 2006). For many researchers, the job has been an issue of concern, mainly because of its connection with turnover, absenteeism, and performance (Iaffaldano and Muchinsky, 1985; Tzeng, 2002; Wegge et al., 2007). There are different factors that influence job satisfaction, although we can indicate that the employees and the nature of work itself are the two main factors (Glisson and Durick, 1988). Igbaria and Guimaraes (1993) noted that professional workers' characteristics and motivation factors differ from those of general employees. Employees should be assigned the jobs with different characteristics that match their individual personality traits. This would encourage them and allow them to experience higher job satisfaction.

Hi-tech professional's tasks are difficult and complicated because he/she must effectively perform the IT function and use IT resources to achieve top performance and business value in support of the business strategies of the enterprise. Therefore, motivation for achievement and job satisfaction are positively related for IT professional worker (Yasin, 1996).

There is a considerable debate among academics and business practitioners about key factors of existence and development of organizations and employees; of which the important one is the value of the use of social networking by organizational members. Since working in the field of IT and high technology is challenging, and the development cycle is changing rapidly, the satisfaction of employees at the workplace is very important for company success. Job satisfaction is also an important factor that shapes the future career of the individual. In this chapter, the characteristics of employee satisfaction are shown between the four countries covered by the survey in the field of IT.

4.2   THEORETICAL BACKGROUND

4.2.1   JOB SATISFACTION

Successful organizations recognize that their employees are the primary source of achieving productivity and performance and, therefore, also care about their satisfaction. Job satisfaction has been studied both as a consequence of many individual and work environment characteristics and as an antecedent to many outcomes (Chahal et al., 2013). Lease (1998) noted that employees with higher job satisfaction are usually less absent, less likely to leave, more productive, more likely to display organizational commitment, and more likely to be satisfied with their lives. The concept of job satisfaction is multifaceted and can be defined in various ways. Job satisfaction can be defined as a positive emotional state resulting from the appraisal of one's job or job experiences Locke and Lathan (1990). Hackman and Oldham (1975)'s job characteristic model (JCM) has been used to determine the level of job satisfaction and focuses on five core job dimensions: (1) skill variety, (2) task identity, (3) task significance, (4) autonomy, and (5) feedback, which in turn influence three key psychological states: experienced meaningfulness of the work, experienced responsibility for the work, and knowledge of results.

Authors like Jayasuriya et al. (2012) and Ezeja et al. (2010) define job satisfaction as pleasant and positive emotional reaction to an individual's perception of his/her work, and it is important, particularly, in internal perception of an individual's values and their relationship to the perception of current working conditions. Job satisfaction can be defined as an emotional response in the working environment (Armstrong, 2006). Armstrong (2006) points out that the term job satisfaction refers only to the attitudes and emotions of individuals regardless of their work. This means that a positive and favorable attitude towards their own workplace reflects an individual's job satisfaction, and on the other hand, negative and unfavorable attitude towards one's own workplace reflects an individual's dissatisfaction in the workplace.

Griffin and Bateman (1986) as well as Hackman and Oldman (1975) conceptualize the term as one's emotional reactions to his/her job. Guimaraes et al. (1992) define job satisfaction as the primary affective response of employees to different aspects of the job and its experiences. Hackman and Oldham (1975) explained that if a job satisfied the core job characteristics the employee would perceive that the work was worthwhile, would feel responsible for the work and would know if the work had been completed satisfactorily or not. The outcome of this would be high-quality work performance and high job satisfaction as a result of intrinsic motivation (Armstrong, 2014). Different approach to job satisfaction is offered by newer models including time and dynamics as important factors influencing the current job satisfaction.

Another, more recent view is to see job satisfaction as a dynamic process, where it arises from a comparison between the current work situation and the aspiration level (Bruggemann et al., 1975; Büssing and Bissels, 1998; Jiménez, 2006; Jiménez et al., 2015).

Work satisfaction or dissatisfaction arises from a comparison between the current work situation and the aspiration level, and the result of this comparison can lead to different forms of job satisfaction (progressive, stabilized, resigned, pseudo, fixated, and constructive job satisfaction) (Bruggemann et al., 1975; Büssing and Bissels, 1998; Büssing et al., 1999). Job satisfaction can also be considered as an emotional response in the working context. As a pleasant feeling the individual perceives the fulfillment of his/her expectations related to work, as determined by the satisfaction of Hollenbeck and Wright (1994). The authors assume three important aspects of employee satisfaction in the workplace:

1. Value: Job satisfaction is a function of the value, which an individual seeks to achieve.
2. Relevance: It is important for realization of the individual values from the perspective of the individual.
3. Detection means detecting the current situation from the perspective of the individual, and comparing it with values.

Beggs (2010) states that the organizational rules, the nature of work assignments, and direct feedback to employees are closely associated with increased satisfaction in the workplace. Similarly, Sultan (2012) points out that, in general, employee satisfaction in the workplace is also affected by challenges in the workplace, autonomy of labor, and different types of work (Armstrong, 2006). Fischer et al. (2013) define that job satisfaction as an emotional state can be regulated downwards or upwards and that this emotional regulation again influences the subjective job satisfaction. In this view, job satisfaction is a dynamic process (Büssing and Bissels, 1998; Jiménez, 2006; Jiménez et al., 2015). Job satisfaction is a result of individuals' perception and evaluation of their job influenced by their own unique needs, values, and expectations, which they regard as being important to them (Sempane et al., 2002). However, if job satisfaction is absent and other work opportunities present themselves, turnover could increase thoroughly (Martins and Coetzee, 2007).

There is also an important connection between retention and job satisfaction. Cropanzano et al., 1997 found out that job satisfaction levels of employees (former colleagues) who stayed at work were affected negatively due to the stress of adjusting to new coworkers. When talking about IT workers' job satisfaction, many authors said that these employees tend to focus on different job satisfaction factors than the ones that satisfy other kinds of employees: autonomy in their work and having opportunities to advance (Glen, 2003).

As discussed earlier, job satisfaction is a main driver of turnover among hi-tech professionals. The strategic value of these professionals in facilitating a smooth and reliable flow of information in organizations, and even providing the basic infrastructure for many new business models, has rendered their turnover.

4.2.2 RETENTION OF HR POLICIES FOR HI-TECH PROFESSIONALS

With the connection to job satisfaction we have to mention talent retention that has become a major concern for all organizations. Hi-tech professional's intentions to leave a company will increase if the HRD does not use all means to develop these employees and to fulfill their ambitions. Certainly, the development is an important factor that also influences the employee satisfaction at the workplace, because it defines a way the employees experience the work and influences people's attitude towards their jobs. Recruitment doesn't happen in an ordinary way; the hi-tech professionals are recruited trough head-hunting agency or directly by another potential employer and, therefore, their job mobility is increasing, so the organizations are finding it more difficult to retain these employees. Therefore, the retention policies in the organization are an important resource to retain this professional in the company, especially is the retention critically important where financial sustainability and survival depend on scarce human and specialist skills such as hi-tech skills (Pienaar and Bester, 2008). It is important to mention that these highly skilled employees have high level of mobility, because the psychological contract has shifted from a previous emphasis on job security and loyalty to the current emphasis on employability and loyalty to own career and experience (Sutherland, 2005).

There is a common challenge how should companies approach retention and retain their high-tech employees. Proactive, not reactive retention policies are the ones that are important. Retention issues that are ignored until the company suspects an employee might leave the company are reactive retention policies. The company uses the retention strategy only when it reaches to the point where it addresses that a hi-tech employee may leave and therefore it reactively offers the employee some kind of enticement to stay. If they succeed the employee will be back to business as usual. This approach might work in the short-term, but does not cultivate longer-term loyalty and job satisfaction. As mentioned above, a better approach is to address retention policy proactively, as a strategic issue. Every employee is motivated by different things and, thus, the retention strategies need to be tailored to the employee's individual needs.

In high-tech firms, employees are often required to have a relatively broad skill range and to coordinate activities across departments to accomplish tasks. The work motivation of hi-tech professional influences: (1) self-perceived task attainment, (2) interest, and (3) utility. Therefore, the employees with higher levels of attainment value, interest, and utility will be more willing to devote greater effort to performing jobs, enjoy their job content more, and will have a greater likelihood of achieving their personal goals (Jung-Yu et al., 2010). Consequently, higher perception of task value will lead to higher work motivation and then improved job satisfaction. The extrinsic factors, such as a competitive salary, good interpersonal relationships, friendly working environment, and job security are key motivational variables that influence employee retention (Samuel and Chipunza, 2009). Therefore, the combination of both intrinsic and extrinsic variables is the most important (Samuel and Chipunza, 2009). One's retention in the company can be positively affected by the employee engagement (Cook and Green, 2011).

4.3 DIFFERENCE BETWEEN JOB SATISFACTION IN RESEARCHED COUNTRIES

There are three important dimensions of job satisfaction that summarize the most important aspects of job satisfaction:

1. Job satisfaction is an emotional response to a job situation.
2. Job satisfaction is determined by the employee perception of how well their outcomes meet or exceed their expectations.
3. Job satisfaction represents several related attitudes, which belong to the most important characteristics of job such as: the work itself, pay, promotion opportunities, supervision, and coworkers (Tella et al., 2007).

Job satisfaction differs across countries—this was pointed out in the research of Sousa-Poza and Sousa-Poza (2000), where they have analyzed 21 countries regarding their job satisfaction. Their research pointed out that countries with high work-role output (e.g., high income, high job security, high latitude) have higher ranking of job satisfaction such as Denmark, Cyprus, Switzerland, Israel, or the Netherlands. The study result of De Witte and Näswall (2003) where the job satisfaction of four countries (Belgium, Netherlands, Italy, and Sweden) was investigated showed that job insecurity and satisfaction are negatively related in all countries. According to these results, the economic status of a country might negatively influence job satisfaction of employees.

4.3.1 FOUR COUNTRIES MACROECONOMIC CHARACTERISTICS

Job satisfaction was researched in four countries: Austria, Germany, Slovenia, and Spain. The economic situations of the countries in question differ, especially between Slovenia, Spain, and the Austria/German cluster, which can be observed in Table 4.1.

TABLE 4.1 Macroeconomic Factors for Slovenia, Spain, Austria, and Germany in 2014. (*Source*: Eurostat (2015)

	Austria	Germany	Slovenia	Spain	EU (28)
Employment rate by age group (20–64) in %	74.2	77.7	67.7	59.9	69.2
Persons employed part time in %	27.9	27.6	11.2	15.9	20.4
Unemployment age, age 25–74 years	5.6	5.0	9.7	24.5	10.2
Long-term unemployment rate, age 25–74 years in %	1.5	2.2	5.3	12.9	5.1
Youth unemployed (under 25 years)	10.3	7.7	20.2	53.2	22.2
Life satisfaction	8.4	7.6	7.8	7.4	7.6
Job satisfaction	8.2	7.3	7.8	7.0	7.2

In Germany, employment rate by age group and a low unemployment rate are the highest in all categories. In Austria, the share of persons working part time and long-term unemployed people is the lowest. In Slovenia, the percent of persons working part time is the lowest and unemployment rate and long-term unemployment rates are relatively high compared to Austria and Germany. In Spain, unemployment rate, long-term unemployment rate, and youth unemployment are the highest among all countries. Austria is leading by having the highest value of job and life satisfaction, followed by Slovenia and Germany.

4.4 RESEARCH

The research question we explored was: Are there any differences in job satisfaction regarding industry in four countries, Germany, Austria, Slovenia, and Spain? Therefore, our main hypothesis is: Differences in job satisfaction regarding industry in four countries, Germany, Austria, Slovenia, and Spain, are statistically significant.

To analyze our research question and to verify hypothesis, we calculated analysis of variance (ANOVA) and post hoc tests to check the differences between industries of the four countries.

4.4.1   MEASURING INSTRUMENT

Job satisfaction was measured with the profile analysis of job satisfaction (PAJS, Jiménez, 2008). The PAJS assesses 11 facets of job satisfaction with 3–5 items (the form of 39 items was used in this study). The items can be answered on a 5-Point-Likert-scale ranging from 1 (satisfied) to 5 (dissatisfied). In this sense, lower scores refer to higher job satisfaction. Studies show that the 11 facets reach a Cronbach Alpha from 0.82 to 0.91. The collected data were processed with the Statistical Package for Social Sciences (SPSS; Version 21). As the variables are interval-scaled, the independent samples ANOVA test was used.

4.4.2   PROCEDURE

A quantitative survey within a cross-cultural project was performed to collect data[1]. It was derived from the online study in the period of 9–19 July, 2013. In order to be eligible for the survey, the participants would have to have a job of at least 10 working hours per week and not to work alone. The participants were categorized as either leaders or employees.

4.4.3   DATA COLLECTING AND RESPONDENTS

In 2013, the total of 2407 employees from Austria, Germany, Slovenia, and Spain were invited to complete the online questionnaire. At the end, 1207 questionnaires were completed; 294 respondents were from Austria, 261 from Spain, 369 from Slovenia, and 283 from Germany. The respondents in each country were from different organizations and some characteristics are presented in Tables 4.2–4.5. The sample is indirectly determined with the project budget "grenz-frei."

The largest proportion of respondents, almost 28%, were from Slovenia, followed by 24.5% from Austria and Germany, and 23.4% were from Spain. Table 4.2 above shows that more than 54% were female respondents and slightly less than 46% were male respondents among all surveyed.

---

[1] Data were collected within the project "culture4leadership," funded by the Province of Styria, Austria, within the framework of "grenz-frei" (i.e., no border) project.

TABLE 4.2 Respondents Regarding Country.

Country		Frequency	Percent	Valid percent	Cumulative percent
Valid	Austria	258	24.5	24.5	24.5
	Spain	247	23.4	23.4	47.9
	Slovenia	292	27.7	27.7	75.5
	Germany	258	24.5	24.5	100
	Total	1055	100	100	

TABLE 4.3 Respondents Regarding Sex.

Sex		Frequency	Percent	Valid percent	Cumulative percent
Valid	Male	482	45.7	45.7	45.7
	Female	573	54.3	54.3	100
	Total	1055	100	100	

TABLE 4.4 Respondents Regarding Age.

Age					
Age in years		Frequency	Percent	Valid percent	Cumulative percent
Valid	Until 20	14	1.3	1.3	1.3
	21–25	58	5.5	5.5	6.8
	26–30	139	13.2	13.2	20
	31–35	178	16.9	16.9	36.9
	36–40	165	15.6	15.6	52.5
	41–45	142	13.5	13.5	66
	46–50	125	11.8	11.8	77.8
	51–55	112	10.6	10.6	88.4
	56–60	81	7.7	7.7	96.1
	61–65	38	3.6	3.6	99.7
	66 years and older	3	0.3	0.3	100
	Total	1055	100	100	

Most of the respondents were in the age group from 31 to 45 years of age (46% of all respondents).

TABLE 4.5   Respondents Regarding Education.

Highest education		Frequency	Percent	Valid percent	Cumulative percent
Valid	Compulsory education	20	1.9	1.9	1.9
	Apprenticeship	281	26.6	26.6	28.5
	Finished grammar/ secondary school/college	250	23.7	23.7	52.2
	University	504	47.8	47.8	100
	Total	1055	100	100	

University degree had 47.8% of all respondents, followed by apprenticeship (26.6%).

4.5   RESULTS

We have used quantitative methods (descriptive statistics) to do research for job satisfaction of employees in four countries. Results of descriptive statistics are presented in Table 4.2. Based on this table we can conclude that:

1. Satisfaction with…working conditions (e.g., flexible working time, occupational health management, …) has highest mean value in commerce/trades (3.26).
2. Satisfaction with…Chances for making a career has highest mean value in public sector (3.20).
3. Satisfaction with…Payment has highest mean value in commerce/ trades (3.12).
4. Satisfaction with…Information and communication has highest mean value in public sector (3.11).
5. Satisfaction with…Organization and management of the company has highest mean value in public sector (3.11).
6. Satisfaction with…compatibility of private and business life has highest mean value in gastronomy and tourism (2.93).
7. Satisfaction with…benefits (e.g., financial, social, security, …) has highest mean value in commerce/trades (2.93).
8. Satisfaction with…Conditions at my working place has highest mean value in public sector (2.84).
9. Satisfaction with…Possibility of making free decisions has highest mean value in commerce/trades (2.79).

10. Satisfaction with...Relationship with your nearest boss has highest mean value in gastronomy and tourism (2.77).
11. Satisfaction with...Regulation of my working time and how I can take my holidays has highest mean value in commerce/trades (2.76).
12. Satisfaction with...Having a demanding job has highest mean value in IT (2.70).
13. Satisfaction with...privacy has highest mean value in commerce/trades (2.69).
14. Satisfaction with...the working task itself has highest mean value in IT (2.64).
15. All in all, satisfaction with my work...has highest mean value in IT (2.62).
16. Satisfaction with...Relationship with your closest colleagues has highest mean value in craftsmanship and building/construction (2.53).

If we look at the average value of individual items according of the industry, then we can see that IT achieves above-average value for all items except for one (Satisfaction with...Relationship with your closest colleagues).

Table 4.7 shows that job satisfaction significantly differs among industries in some cases ($p < 0.5$).

In Table 4.7, we can see the calculated statistical significance of p that is greater 0.05 at the followings:

- Satisfaction with... Information and communication;
- Satisfaction with... Conditions at my working place;
- Satisfaction with... Payment;
- Satisfaction with... Regulation of my working time and how I can take my holidays;
- Satisfaction with... working conditions (e.g., flexible working time, occupational health management, ...);
- Satisfaction with... benefits (e.g., financial, social, security...);
- All in all satisfaction with my work;
- Satisfaction with... compatibility of private and business life;
- Satisfaction with... Chances for making a career;
- Satisfaction with... Organization and management of the company;

TABLE 4.6 Descriptive Statistics for Job Satisfaction with Regarding Industry of Respondents of Four Countries. (*Source:* Authors)

		N	Mean	Std. deviation	Std. error	95% confidence interval for mean	
						Lower bound	Upper bound
Satisfaction with…information and communication	Public sector	100	3.11	1.063	0.106	2.90	3.32
	Industry	115	2.92	1.148	0.107	2.71	3.13
	Health care sector	65	2.91	1.182	0.147	2.61	3.20
	Commerce/ trades	68	2.96	1.028	0.125	2.71	3.20
	Education	72	2.72	0.967	0.114	2.49	2.95
	Gastronomy and tourism	30	2.63	1.299	0.237	2.15	3.12
	Credit and insurance	39	2.79	1.005	0.161	2.47	3.12
	Craftsmanship and building/ construction	36	2.81	1.142	0.190	2.42	3.19
	IT	89	3.04	1.177	0.125	2.80	3.29
	Other services	67	2.78	1.126	0.138	2.50	3.05
	Logistics/ transport	40	2.95	1.037	0.164	2.62	3.28
	Other	185	2.85	1.108	0.081	2.69	3.01
	999	38	2.61	0.974	0.158	2.29	2.93
	Total	944	2.88	1.103	0.036	2.81	2.95

TABLE 4.6 *(Continued)*

		N	Mean	Std. devia-tion	Std. error	95% confidence interval for mean	
						Lower bound	Upper bound
Satisfaction with... having a demanding job	Public sector	100	2.62	1.013	0.101	2.42	2.82
	Industry	115	2.54	1.078	0.101	2.34	2.74
	Health care sector	65	2.23	0.862	0.107	2.02	2.44
	Commerce/ trades	68	2.51	1.015	0.123	2.27	2.76
	Education	72	2.21	0.804	0.095	2.02	2.40
	Gastronomy and tourism	30	2.40	1.102	0.201	1.99	2.81
	Credit and insurance	39	2.23	0.931	0.149	1.93	2.53
	Craftsmanship and building/ construction	36	2.17	1.082	0.180	1.80	2.53
	IT	89	2.70	1.102	0.117	2.46	2.93
	Other services	67	2.42	0.940	0.115	2.19	2.65
	Logistics/ transport	40	2.40	0.900	0.142	2.11	2.69
	Other	185	2.49	0.973	0.072	2.35	2.63
	999	38	2.55	1.005	0.163	2.22	2.88
	Total	944	2.46	0.996	0.032	2.40	2.52

TABLE 4.6   *(Continued)*

		N	Mean	Std. devia-tion	Std. error	95% confidence interval for mean	
						Lower bound	Upper bound
Satisfaction with… rela-tionship with your closest colleagues	Public sector	100	2.40	1.110	0.111	2.18	2.62
	Industry	115	2.15	0.948	0.088	1.97	2.32
	Health care sector	65	2.28	1.053	0.131	2.02	2.54
	Commerce/trades	68	2.26	0.924	0.112	2.04	2.49
	Education	72	2.03	0.855	0.101	1.83	2.23
	Gastronomy and tourism	30	2.27	0.907	0.166	1.93	2.61
	Credit and insurance	39	2.10	0.912	0.146	1.81	2.40
	Craftsmanship and building/construction	36	2.53	1.000	0.167	2.19	2.87
	IT	89	2.18	1.018	0.108	1.97	2.39
	Other services	67	2.03	.870	0.106	1.82	2.24
	Logistics/transport	40	2.53	1V219	0.193	2.14	2.91
	Other	185	2.34	1.061	0.078	2.18	2.49
	999	38	2.26	0.891	0.145	1.97	2.56
	Total	944	2.25	1.003	0.033	2.19	2.32

TABLE 4.6 *(Continued)*

		N	Mean	Std. deviation	Std. error	95% confidence interval for mean	
						Lower bound	Upper bound
Satisfaction with… relationship with your nearest boss	Public sector	100	2.76	1.280	0.128	2.51	3.01
	Industry	115	2.61	1.168	0.109	2.39	2.82
	Health care sector	65	2.72	1.269	0.157	2.41	3.04
	Commerce/ trades	68	2.74	1.205	0.146	2.44	3.03
	Education	72	2.39	1.170	0.138	2.11	2V66
	Gastronomy and tourism	30	2.77	1.194	0.218	2.32	3.21
	Credit and insurance	39	2.23	1.087	0.174	1.88	2.58
	Craftsmanship and building/ construction	36	2.78	1.290	0.215	2.34	3.21
	IT	89	2.69	1.249	0.132	2.42	2.95
	Other services	67	2.49	1.146	0.140	2.21	2.77
	Logistics/ transport	40	2.75	1.296	0.205	2.34	3.16
	Other	185	2.68	1.171	0.086	2.51	2.85
	999	38	2.89	1.085	0.176	2.54	3.25
	Total	944	2.65	1.203	0.039	2.57	2.73

TABLE 4.6 *(Continued)*

		N	**Mean**	**Std. deviation**	**Std. error**	**95% confidence interval for mean**	
						Lower bound	**Upper bound**
Satisfaction with… organization and management of the company	Public sector	100	3.11	1.081	0.108	2.90	3.32
	Industry	115	2.79	1.064	0.099	2.59	2.99
	Health care sector	65	2.77	1.101	0.137	2.50	3.04
	Commerce/ trades	68	2.85	1.213	0.147	2.56	3.15
	Education	72	2.61	1.029	0.121	2.37	2.85
	Gastronomy and tourism	30	2.50	0.974	0.178	2.14	2.86
	Credit and insurance	39	2.62	0.963	0.154	2.30	2.93
	Craftsmanship and building/ construction	36	2.72	1.186	0.198	2.32	3.12
	IT	89	2.93	1.204	0.128	2.68	3.19
	Other services	67	2.69	1.183	0.145	2.40	2.98
	Logistics/ transport	40	2.90	1.215	0.192	2.51	3.29
	Other	185	2.83	1.075	0.079	2.67	2.98
	999	38	2.79	1.094	0.178	2.43	3.15
	Total	944	2.81	1.110	0.036	2.74	2.88

TABLE 4.6 *(Continued)*

		N	Mean	Std. devia-tion	Std. error	95% confidence interval for mean	
						Lower bound	Upper bound
Satisfaction with... chances for making a career	Public sector	100	3.20	1.155	0.115	2.97	3.43
	Industry	115	2.89	1.041	0.097	2.69	3.08
	Health care sector	65	2.88	1.083	0.134	2.61	3.15
	Commerce/ trades	68	3.06	1.170	0.142	2.78	3.34
	Education	72	2.96	1.027	0.121	2.72	3.20
	Gastronomy and tourism	30	3.00	1.390	0.254	2.48	3.52
	Credit and insurance	39	2.74	1.093	0.175	2.39	3.10
	Craftsmanship and building/ construction	36	3.00	1.373	0.229	2.54	3.46
	IT	89	3.16	1.176	0.125	2.91	3.41
	Other services	67	2.73	1.162	0.142	2.45	3.01
	Logistics/ transport	40	2.98	1.000	0.158	2.66	3.29
	Other	185	3.06	1.159	0.085	2.90	3.23
	999	38	2.95	1.184	0.192	2.56	3.34
	Total	944	2.99	1.143	0.037	2.92	3.07

TABLE 4.6   *(Continued)*

		N	Mean	Std. devia-tion	Std. error	95% confidence interval for mean	
						Lower bound	Upper bound
Satisfaction with… condi-tions at my working place	Public sector	100	2.84	1.135	0.113	2.61	3.07
	Industry	115	2.54	1.028	0.096	2.35	2.73
	Health care sector	65	2.46	0.937	0.116	2.23	2.69
	Commerce/ trades	68	2.72	1.091	0.132	2.46	2.98
	Education	72	2.57	0.932	0.110	2.35	2.79
	Gastronomy and tourism	30	2.47	1.196	0.218	2.02	2.91
	Credit and insurance	39	2.23	0.931	0.149	1.93	2.53
	Craftsmanship and building/ construction	36	2.56	1.182	0.197	2.16	2.96
	IT	89	2.70	1.102	0.117	2.46	2.93
	Other services	67	2.31	0.988	0.121	2.07	2.55
	Logistics/ transport	40	2.43	0.931	0.147	2.13	2.72
	Other	185	2.61	1.185	0.087	2.43	2.78
	999	38	2.66	0.938	0.152	2.35	2.97
	Total	944	2.58	1.073	0.035	2.51	2.65

TABLE 4.6 *(Continued)*

		N	Mean	Std. devia-tion	Std. error	95% confidence interval for mean	
						Lower bound	Upper bound
Satisfaction with... possi-bility of making free decisions	Public sector	100	2.76	1.065	0.106	2.55	2.97
	Industry	115	2.57	1.026	0.096	2.38	2.76
	Health care sector	65	2.54	1.032	0.128	2.28	2.79
	Commerce/ trades	68	2.79	1.153	0.140	2.51	3.07
	Education	72	2.31	0.929	0.109	2.09	2.52
	Gastronomy and tourism	30	2.60	1.192	0.218	2.15	3.05
	Credit and insurance	39	2.36	1.063	0.170	2.01	2.70
	Craftsmanship and building/ construction	36	2.19	1.142	0.190	1.81	2.58
	IT	89	2.69	1.040	0.110	2.47	2.90
	Other services	67	2.37	1.126	0.138	2.10	2.65
	Logistics/ transport	40	2.60	1.008	0.159	2.28	2.92
	Other	185	2.57	1.169	0.086	2.40	2.74
	999	38	2.82	1.036	0.168	2.48	3.16
	Total	944	2.57	1.088	0.035	2.50	2.64

TABLE 4.6 *(Continued)*

		N	Mean	Std. devia-tion	Std. error	95% confidence interval for mean	
						Lower bound	Upper bound
Satisfaction with... privacy	Public sector	100	2.55	1.019	0.102	2.35	2.75
	Industry	115	2.41	1.042	0.097	2.22	2.60
	Health care sector	65	2.31	1.103	0.137	2.03	2.58
	Commerce/ trades	68	2.69	1.083	0.131	2.43	2.95
	Education	72	2.25	0.975	0.115	2.02	2.48
	Gastronomy and tourism	30	2.40	1.102	0.201	1.99	2.81
	Credit and insurance	39	2.03	0.778	0.125	1.77	2.28
	Craftsmanship and building/ construction	36	2.22	0.929	0.155	1.91	2.54
	IT	89	2.64	1.160	0.123	2.40	2.88
	Other services	67	2.30	1.101	0.135	2.03	2.57
	Logistics/ transport	40	2.45	1.037	0.164	2.12	2.78
	Other	185	2.45	1.078	0.079	2.29	2.60
	999	38	2.66	0.966	0.157	2.34	2.98
	Total	944	2.44	1.056	0.034	2.37	2.50

TABLE 4.6 *(Continued)*

		N	Mean	Std. devia-tion	Std. error	95% confidence interval for mean	
						Lower bound	Upper bound
Satisfaction with... payment	Public sector	100	3.07	1.157	0.116	2.84	3.30
	Industry	115	2.84	1.097	0.102	2.64	3.05
	Health care sector	65	2.78	1.179	0.146	2.49	3.08
	Commerce/ trades	68	3.12	1.216	0.147	2.82	3.41
	Education	72	2.94	0.902	0.106	2.73	3.16
	Gastronomy and tourism	30	2.93	1.363	0.249	2.42	3.44
	Credit and insurance	39	2.54	1.072	0.172	2.19	2.89
	Craftsmanship and building/ construction	36	2.61	0.994	0.166	2.27	2.95
	IT	89	2.82	1.134	0.120	2.58	3.06
	Other services	67	2.87	1.266	0.155	2.56	3.17
	Logistics/ transport	40	2.73	1.198	0.189	2.34	3.11
	Other	185	2.85	1.163	0.086	2.69	3.02
	999	38	2.74	1.057	0.172	2.39	3.08
	Total	944	2.86	1.142	0.037	2.79	2.94

TABLE 4.6 *(Continued)*

		N	Mean	Std. deviation	Std. error	95% confidence interval for mean	
						Lower bound	Upper bound
Satisfaction with... regulation of my working time and how I can take my holidays.	Public sector	100	2.44	1.008	0.101	2.24	2.64
	Industry	115	2.48	1.119	0.104	2.27	2.68
	Health care sector	65	2.37	1.167	0.145	2.08	2.66
	Commerce/ trades	68	2.76	1.259	0.153	2.46	3.07
	Education	72	2.33	0.904	0.107	2.12	2.55
	Gastronomy and tourism	30	2.60	1.248	0.228	2.13	3.07
	Credit and insurance	39	2.13	0.894	0.143	1.84	2.42
	Craftsmanship and building/ construction	36	2.39	0.964	0.161	2.06	2.72
	IT	89	2.48	1.088	0.115	2.25	2.71
	Other services	67	2.45	1.049	0.128	2.19	2.70
	Logistics/ transport	40	2.60	1.172	0.185	2.23	2.97
	Other	185	2.60	1.162	0.085	2.43	2.77
	999	38	2.68	1.042	0.169	2.34	3.03
	Total	944	2.50	1.100	0.036	2.43	2.57

TABLE 4.6 *(Continued)*

		N	Mean	Std. devia-tion	Std. error	95% confidence interval for mean	
						Lower bound	Upper bound
Satisfaction with… working conditions (e.g., flexible working time. occupa-tional health management. …)	Public sector	100	3.01	1,133	0.113	2.79	3.23
	Industry	115	2.79	1,072	0.100	2.59	2.99
	Health care sector	65	2.85	1.149	0.142	2.56	3.13
	Commerce/ trades	68	3.26	1.167	0.142	2.98	3.55
	Education	72	2.93	0.998	0.118	2.70	3.16
	Gastronomy and tourism	30	3.10	1.348	0.246	2.60	3.60
	Credit and insurance	39	2.51	0.997	0.160	2.19	2. .84
	Craftsmanship and building/ construction	36	2.94	1.145	0.191	2.56	3.33
	IT	89	3.02	1.044	0.111	2.80	3.24
	Other services	67	2.96	1.236	0.151	2.65	3.26
	Logistics/ transport	40	2.93	1.071	0.169	2.58	3.27
	Other	185	2.96	1.177	0.087	2.79	3.13
	999	38	2.87	0.991	0.161	2.54	3.19
	Total	944	2.94	1.125	0.037	2.87	3.01

TABLE 4.6 *(Continued)*

		N	Mean	Std. devia- tion	Std. error	95% confidence interval for mean	
						Lower bound	Upper bound
Satisfaction with... compat- ibility of private and business life	Public sector	100	2.59	1.065	0.106	2.38	2.80
	Industry	115	2.57	1.093	0.102	2.37	2.78
	Health care sector	65	2.54	1.238	0.154	2.23	2.85
	Commerce/ trades	68	2.85	1.307	0.159	2.54	3.17
	Education	72	2.49	1.007	0.119	2.25	2.72
	Gastronomy and tourism	30	2.93	1.337	0.244	2.43	3.43
	Credit and insurance	39	2.36	0.873	0.140	2.08	2.64
	Craftsmanship and building/ construction	36	2.75	1.025	0.171	2.40	3.10
	IT	89	2.73	1.074	0.114	2.50	2.96
	Other services	67	2.58	1.130	0.138	2.31	2.86
	Logistics/ transport	40	2.75	1.080	0.171	2.40	3.10
	Other	185	2.74	1.169	0.086	2.57	2.91
	999	38	2.61	0.823	0.134	2.33	2.88
	Total	944	2.65	1.116	0.036	2.58	2.72

TABLE 4.6 *(Continued)*

		N	Mean	Std. devia-tion	Std. error	95% confidence interval for mean	
						Lower bound	Upper bound
Satisfaction with... the working task itself	Public sector	100	2.59	0.954	0.095	2.40	2.78
	Industry	115	2.41	0.926	0.086	2.24	2.58
	Health care sector	65	2.17	1.054	0.131	1.91	2.43
	Commerce/ trades	68	2.62	1.093	0.133	2.35	2.88
	Education	72	2.25	0.818	0.096	2.06	2.44
	Gastronomy and tourism	30	2.20	0.887	0.162	1.87	2.53
	Credit and insurance	39	2.18	0.970	0.155	1.87	2.49
	Craftsmanship and building/ construction	36	2.14	0.961	0.160	1.81	2.46
	IT	89	2.64	1.003	0.106	2.43	2.85
	Other services	67	2.28	0.901	0.110	2.06	2.50
	Logistics/ transport	40	2.30	0.791	0.125	2.05	2.55
	Other	185	2.42	0.947	0.070	2.28	2.56
	999	38	2.68	0.933	0.151	2.38	2.99
	Total	944	2.41	0.960	0.031	2.35	2.47

TABLE 4.6   *(Continued)*

		N	Mean	Std. devia-tion	Std. error	95% confidence interval for mean	
						Lower bound	Upper bound
Satisfaction with... benefits (e.g. financial, social, security,)	Public sector	100	2.84	1.117	0.112	2.62	3.06
	Industry	115	2.57	0.974	0.091	2.39	2.75
	Health care sector	65	2.63	1.126	0.140	2.35	2.91
	Commerce/ trades	68	2.93	1.188	0.144	2.64	3.21
	Education	72	2.64	1.011	0.119	2.40	2.88
	Gastronomy and tourism	30	2.73	1.337	0.244	2.23	3.23
	Credit and insurance	39	2.44	1.119	0.179	2.07	2.80
	Craftsmanship and building/ construction	36	2.75	1.025	0.171	2.40	3.10
	IT	89	2.88	1.185	0.126	2.63	3.13
	Other services	67	2.72	1.056	0.129	2.46	2.97
	Logistics/ transport	40	2.63	1.102	0.174	2.27	2.98
	Other	185	2.69	1.098	0.081	2.53	2.85
	999	38	2.79	0.963	0.156	2.47	3.11
	Total	944	2.71	1.095	0.036	2.64	2.78

TABLE 4.6 *(Continued)*

		N	Mean	Std. devia- tion	Std. error	95% confidence interval for mean	
						Lower bound	Upper bound
All-in-all, satis- faction with my work…	Public sector	100	2.51	1.087	0.109	2.29	2.73
	Industry	115	2.39	0.915	0.085	2.22	2.56
	Health care sector	65	2.32	1.002	0.124	2.07	2.57
	Commerce/ trades	68	2.57	1.055	0.128	2.32	2.83
	Education	72	2.19	0.762	0.090	2.02	2.37
	Gastronomy and tourism	30	2.43	1.135	0.207	2.01	2.86
	Credit and insurance	39	2.23	0.872	0.140	1.95	2.51
	Craftsmanship and building/ construction	36	2.44	1.054	0.176	2.09	2.80
	IT	89	2.62	1.017	0.108	2.40	2.83
	Other services	67	2.54	1.064	0.130	2.28	2.80
	Logistics/ transport	40	2.45	1.061	0.168	2.11	2.79
	Other	185	2.45	1.068	0.078	2.29	2.60
	999	38	2.68	0.962	0.156	2.37	3.00
	Total	944	2.45	1.012	0.033	2.39	2.52

Likert Scale: 1-very satisfied; 2-satisfied; 3-neither satisfied nor dissatisfied; 4-dissatisfied; and 5-very dissatisfied; *IT:* information technology.

- Satisfaction with… Relationship with your closest colleagues;
- Satisfaction with… Relationship with your nearest boss.

At these items, therefore, no statistically significant differences exist. Statistically significant differences exist in the following items:

- Satisfaction with… privacy;
- Satisfaction with… the working task itself;
- Satisfaction with… Possibility of making free decisions;
- Satisfaction with… Having a demanding job.

Therefore, we performed a post hoc analysis, which is based on the Tukey honest significant difference (HSD) method. The results are presented in Tables 4.4–4.7. We can see that no statistical significant results are presented.

TABLE 4.7   The ANOVA Results for the Differences in the Job Satisfaction According to the Industry in Four Countries. (*Source*: Authors)

		The sum of the squares	df	Average squares	F	p
Satisfaction with… Information and communication	Between groups	16,403	12	1.367	1.126	0.335
	Within groups	1130,012	931	1.214		
	Total	1146,414	943			
Satisfaction with… Having a demanding job	Between groups	22,415	12	1.868	1.903	0.031
	Within groups	913,974	931	982		
	Total	936,389	943			
Satisfaction with… Relationship with your closest colleagues	Between groups	18,759	12	1.563	1.564	0.097
	Within groups	930,740	931	1.000		
	Total	949,499	943			
Satisfaction with… Relationship with your nearest boss	Between groups	19,633	12	1.636	1.133	0.329
	Within groups	1344,705	931	1.444		
	Total	1364,338	943			
Satisfaction with… Organization and management of the company	Between groups	19,505	12	1.625	1.325	0.198
	Within groups	1142,307	931	1.227		
	Total	1161,812	943			
Satisfaction with… Chances for making a career	Between groups	17,299	12	1.442	1.106	0.351
	Within groups	1213,675	931	1.304		
	Total	1230,974	943			
Satisfaction with… Conditions at my working place	Between groups	21,660	12	1.805	1.579	0.092
	Within groups	1064.539	931	1.143		
	Total	1086.199	943			

TABLE 4.7 *(Continued)*

		The sum of the squares	df	Average squares	F	p
Satisfaction with... Possibility of making free decisions	Between groups	25,083	12	2.090	1.785	0.046
	Within groups	1090,303	931	1.171		
	Total	1115,386	943			
Satisfaction with... privacy	Between groups	24,520	12	2.043	1.851	0.037
	Within groups	1027,538	931	1.104		
	Total	1052.058	943			
Satisfaction with... Payment	Between groups	17,698	12	1.475	1.132	0.330
	Within groups	1212,946	931	1.303		
	Total	1230,644	943			
Satisfaction with... Regulation of my working time and how I can take my holidays.	Between groups	18,169	12	1.514	1.257	0.239
	Within groups	1121,826	931	1.205		
	Total	1139,996	943			
Satisfaction with... working conditions (e.g., flexible working time, occupational health management, ...)	Between groups	19,575	12	1.631	1.294	0.216
	Within groups	1173,336	931	1.260		
	Total	1192,911	943			
Satisfaction with... compatibility of private and business life	Between groups	15,510	12	1.293	1.039	0.410
	Within groups	1158,524	931	1.244		
	Total	1174,034	943			
Satisfaction with... the working task itself	Between groups	26,988	12	2.249	2.489	0.003
	Within groups	841,358	931	0,904		
	Total	868,346	943			

TABLE 4.7   *(Continued)*

		The sum of the squares	df	Average squares	F	p
Satisfaction with... benefits (e.g., financial, social, security)	Between groups	13,866	12	1.155	0.963	0.483
	Within groups	1116,909	931	1.200		
	Total	1130,775	943			
All-in-all, satis-faction with my work...	Between groups	14,532	12	1.211	1.185	0.289
	Within groups	951,226	931	1,022		
	Total	965,758	943			

APPENDIX

TABLE 4.8   Results of Tukey honest significant difference (HSD) Method for Item Satisfaction with... Having a Demanding Job. (*Source*: Authors)

Tukey HSD[a,b]		
Industrial sector categorized	*N*	**Subset for alpha =0.05**
		1
Craftsmanship and building/construction	36	2.17
Education	72	2.21
Health care sector	65	2.23
Credit and insurance	39	2.23
Gastronomy and tourism	30	2.40
Logistics/transport	40	2.40
Other services	67	2.42
Other	185	2,49
Commerce/trades	68	2.51
Industry	115	2.54
999	38	2.55
Public Sector	100	2.62
IT	89	2.70
Sig.		0.190

Means for groups in homogeneous subsets are displayed. [a]Uses Harmonic Mean Sample Size = 55.960. [b]The group sizes are unequal. The harmonic mean of the group sizes is used. Type I error levels are not guaranteed.

TABLE 4.9 Results of Tukey HSD Method for Satisfaction with… possibility of Making Free Decisions Source: Authors

Tukey HSD[a,b]		
Industrial sector categorized	*N*	Subset for alpha = 0.05
		1
Craftsmanship and building/construction	36	2.19
Education	72	2.31
Credit and insurance	39	2.36
Other services	67	2.37
Health care sector	65	2.54
Other	185	2.57
Industry	115	2.57
Gastronomy and tourism	30	2.60
Logistics/transport	40	2.60
IT	89	2.69
Public sector	100	2.76
Commerce/trades	68	2.79
999	38	2.82
Sig.		0.113

Means for groups in homogeneous subsets are displayed.[a]Uses Harmonic Mean Sample Size = 55.960. [b]The group sizes are unequal. The harmonic mean of the group sizes is used. Type I error levels are not guaranteed.

TABLE 4.10 Results of Tukey HSD Method for Item Satisfaction with… Privacy. (*Source*: Authors)

Tukey HSD[a,b]			
Industrial sector categorized	*N*	Subset for alpha = 0.05	
		1	**2**
Credit and insurance	39	2.03	
Craftsmanship and building/construction	36	2.22	2.22
Education	72	2.25	2.25
Other services	67	2.30	2.30

TABLE 4.10   *(Continued)*

Tukey HSD[a,b]			
Industrial sector categorized	*N*	**Subset for alpha = 0.05**	
		1	**2**
Health care sector	65	2.31	2.31
Gastronomy and tourism	30	2.40	2.40
Industry	115	2.41	2.41
Other	185	2.45	2.45
Logistics/transport	40	2.45	2.45
Public sector	100	2.55	2.55
IT	89	2.64	2.64
999	38	2.66	2.66
Commerce/trades	68		2.69
Sig.		0.075	0.473

Means for groups in homogeneous subsets are displayed. [a]Uses Harmonic Mean Sample Size = 55.960. [b]The group sizes are unequal. The harmonic mean of the group sizes is used. Type I error levels are not guaranteed.

TABLE 4.11   Results of Tukey HSD Method for Item Satisfaction with… the Working Task Itself. (*Source*: Authors)

Tukey HSD[a,b]		
Industrial sector categorized	*N*	**Subset for alpha = 0.05**
		1
Craftsmanship and building/construction	36	2.14
Health care sector	65	2.17
Credit and insurance	39	2.18
Gastronomy and tourism	30	2.20
Education	72	2.25
Other services	67	2.28
Logistics/transport	40	2.30
Industry	115	2.41
Other	185	2.42

TABLE 4.11   *(Continued)*

Tukey HSD[a,b]		
Industrial sector categorized	N	Subset for alpha = 0.05
		1
Public sector	100	2.59
Commerce/trades	68	2.62
IT	89	2.64
999	38	2.68
Sig.		0.114

Means for groups in homogeneous subsets are displayed. [a]Uses harmonic mean sample size = 55.960. [b]The group sizes are unequal. The harmonic mean of the group sizes is used. Type I error levels are not guaranteed.

4.6   DISCUSSION

In the present study, we investigated cultural differences in employee's job satisfaction regarding industry. In connection to differences in job satisfaction of employees regarding industry, we have formulated our hypothesis that job satisfaction statistically and significantly differs regarding industry. ANOVA and post hoc test showed that there are no statistically significant differences in job satisfaction regarding industry. This means that the job satisfaction is not higher in the IT industry, although the descriptive statistics showed that if we look at the average value of individual items according of the industry, we can see that IT achieves above-average value for all items except for one (Satisfaction with... Relationship with your closest colleagues).

4.7   CONCLUSION

A problem most organizations face is how to retain skilled employees and how to keep their hi-tech professionals satisfied at their job. As technology has gone from common to prevalent, retaining skilled hi-tech professionals has become more difficult, although organizations are paying their knowledge workers higher salaries, increasing bonuses, and so forth. These factors are not important enough to retain these employees in the company. Therefore, the organizations should find the key factors that influence the job satisfaction of their hi-tech workers. Otherwise someone's intention to leave due to one being dissatisfied at the job will also reflect in an organization's bottom line, as the costs associated with discontent employees can readily be measured by looking at what an organization spends hiring and training new workers.

On the other hand, job satisfaction is important to employees because it can affect their general health, happiness, and work–life balance (Greenhause et al., 2002). Dissatisfaction at job means higher rates of absenteeism, employees are more likely to quit their jobs, arrive late for work, produce less than colleagues who are happier in their jobs, and can negatively affect the morale of the organization (Joseph et al., 2007; Reichheld, 1996).

External influences, such as economic insecurity, are considered, but are not specifically highlighted in the feedback model of job satisfaction (Jiménez, 2006). Therefore, we suggest that the theoretical models of job satisfaction should include the environmental factors of the organization and the macroeconomic factors of a national economy, to be more holistic.

The ever-increasing speed that made businesses more dependent on information technologies has turned hi-tech professionals into vital assets for most businesses and also, as mentioned above, their job (dis)satisfaction has been cited as a major reason for their turnover. We believe that a hi-tech professional's job satisfaction has an important role for the company's competitive advantage, because satisfied employees are more engaged and more committed to their work and results. The proactive retention policies have, therefore, been of major importance to retain such professionals in the company. There is also a close relationship between retention policy and job satisfaction, and it presents a fruitful area for further research. The results can also help organizations to better understand the relationship between managers and high-tech professionals and, thereby, increase job satisfaction among one of their most critical profiles in today's highly competitive business environment.

4.8.1   LIMITATIONS AND FURTHER RESEARCH

Future research should highlight the effect of national culture on job satisfaction by analyzing different groups of employees (e.g., gender, age groups, or managers vs. subordinates). It would also be interesting to investigate job satisfaction in all EU countries. Basic limitation is the size of the sample in research.

KEYWORDS

- job satisfaction
- differences
- industry
- IT industry

- Austria
- Germany
- Slovenia
- Spain

REFERENCES

Armstrong, M. *A Handbook of Human Resource Management Practice*, 10th ed.; Kogan Page Publishing: London, 2006; p 264.

Armstrong, M. *A Handbook of Human Resource Management Practice,* 13.th Edition. London: Kogan Page, **2014**.

Bruggemann, A.; Groskurth, P.; Ulich, E. *Arbeitszufriedenheit* (Job Satisfaction). Bern: Hans Huber, **1975**.

Büssing, A.; Bissels, T. Different Forms of Work Satisfaction: Concept and Qualitative Research. *Eur. Psychol.* **1998,** *3*(3), 209–218.

Büssing, A.; Bissels, T.; Fuchs, V.; Perrar, K.-M. A Dynamic Model of work Satisfaction: Qualitative Approaches. *Hum. Relat.* **1999**, *52*(8), 999–1027.

Chahal, A; Chahal, S.; Chowdhary, B.; Chahal, J. Job Satisfaction Among Bank Employees: An Analysis of the Contributing Variables Towards Job Satisfaction. *Int. J. Sci. Technol. Res.* **2013,** *2*(8), 11.

Christen, M.; Iyer, G.; Soberman, D. Job Satisfaction, Job Performance, and Effort: A Reexamination Using Agency Theory. *J. Mark.* **2006,** *70*, 137–150.

Cook, S.; Green, M. A New Approach to Engagement. *Train. J.* **2011**, 23–26.

Cropanzano, R.; Howes, J. C.; Grandey, A. A.; Toth, P. The Relationship of Organizational Politics and Support to Work Behaviors, Attitudes and Stress. *J. Organ. Behav.* **1997,** *18*(2), 159–180.

De Witte, H.; Näswall, K. "Objective" vs "Subjective" Job Insecurity: Consequences of Temporary Work for Job Satisfaction and Organizational Commitment in Four European Countries. *Eco. Ind. Democracy* **2003,** *24*(2), 149–188.

Eurostat. http://ec.europa.eu/eurostat/data/browse-statistics-by-theme. (accessed July 31, 2015).

Ezeja, E. B.; Azodo, C. C.; Ehizele, A. O.; Ehigiator, O.; Oboro, H. O. Assessment of Job Satisfaction and Working Conditions of Nigerian Oral Health Workers. *Int. J. Biomed.*

Health Sci. **2010,** *6(3).* http://www.klobex.org/journals/ijbhs/ijbhs6/ijbhs 630610054. pdf (accessed 15 April, 2016).

Fischer, O.; Fischer, L.; Meyenschein, K. Emotion at Work. *Wirtschaftspsychologie* **2013,** *2/3,* 92–103.

Glen, P. Leading Geeks: How to Manage and Lead People Who Deliver Technology. San Francisco: Jossey-Bass, **2003.**

Glisson, C. V.; Durick, M. Predictors of Job Satisfaction and Organizational Commitment in Human Service Organizations. *Administrative Quarterly* **1988,** *33*(1), 61–68.

Greenhaus, J. H.; Collins, K. M.; Shaw, J. D. The Relation Between Work-Life Balance and Quality of Life. *J. Vocational Behav.* **2002,** *63*(3), 510–531.

Hackman, R. J.; Oldham, G. R. Development of the Job Diagnostic Survey. *J. Appl. Psychol.* **1975,** *60,* 159–170.

Hollenbeck, J., Wright, P. Human Resource Management. Burr Ridgeil: Irwin, **1994.**

Iaffaldano, M. T.; Muchinsky, P. M. Job Satisfaction and Job Performance: A Meta-Analysis. *Psychol. Bull.* **1985,** *97,* 251–273.

Igbaria, M.; Guimaraes, T. Antecedents and Consequences of Job Satisfaction Amount Information Center Employees. *J. Manage. Inf. Syst.* **1993,** *9*(4), 145–174.

Jayasuriya, R.; Whittaker, M.; Halim, G.; Matineau, T. Rural Health Workers and Their Work Environment: The Role of Interpersonal Factors on Job Satisfaction of Nurses in Rural Papua New Guinea. BMC Health Services Research. *J. Bus. Ethics* **2012,** *81*(4), 837–850.

Jiménez, P. Arbeitszufriedenheit als Mittlervariable in Feedbackprozessen. Eine kybernetische Perspektive [Job Satisfaction as Mediator in Feedback Loops. A Cybernetic View]. In *Arbeitszufriedenheit;* L. Fischer (Hrsg) *Konzepte und empirische Befunde,* Göttingen: Hogrefe, **2006;** 160–186.

Jiménez, P. *PAZ Profilanalyse der Arbeitszufriedenheit. Manual Wiener Testsystem (Profile Analysis of Job Satisfaction. Test Manual in Vienna Test System).* Schuhfried GmbH: Mödling, **2008.**

Jiménez, P.; Dunkl, A.; Stolz, R. Anticipation of the Development of Job Satisfaction— Construct and Validation Results of an Indicator for Well-Being at the Workplace. *Psychology* **2015,** *6,* 856–866.

Jung-Yu, L.; Hsin-Jung, C.; Chun-Chieh, Y. Task Value, Goal Orientation, and Employee Job Satisfaction in High-Tech Firms. *Afr. J. Bus. Manage.* **2010,** *5*(1), 76–87.

Joseph, D.; Kok-Yee, N.; Koh, C.; Soon, A. Turnover of Information Technology Professionals: A Narrative Review Meta-Analytic Structural Equation Modeling and Model Development. MIS Quarterly, **2007,** *31*(3), 547–577.

Lease, S. H. Annual Review, 1993–1997: Work Attitudes and Outcomes. *J. Vocational Behav.* **1998,** *53*(2), 154–183.

Locke, E. A; Lathan, G. P. Theory of Goal Setting and Task Performance. Prentice Hall: Englewood Cliffs, NJ, **1990,** 248–250.

Martins, N.; Coetzee, M. Organisational Culture, Employee Satisfaction, Perceived Leader Emotional Compentency and Personality Type: An Exploratory Study in a South African Engineering Company. *South Afr. J. Hum. Res. Manage.* **2007,** *5*(2), 20–32.

Nagy, M. S. Using a Single-item Approach to Measure Facet Job Satisfaction. *J. Occup. Organ. Psychol.* **2002,** *75*(1), 77–86.

Niederman, F.; Sumner, M. Effects of Tasks, Salaries, and Shocks on Job Satisfaction Among MIS Professionals. *Inf. Resour. Manage. J.* **2004,** *17*(4), 49–72.

Pienaar, C.; Bester, C. The Retention of Academics in the Early Career Phase. *SA J. Hum. Resour. Manage.* **2008,** *6*(2), 32–41.

Poulin, J. E. Job Task and Organizational Predicators of Social Worker Job Satisfaction Change: a Panel Study. *Adm. Soc. Work* **1994,** *18*(1), 21–39.

Reichheld, F. F. *The Loyalty Effect: The Hidden Force Behind Growth, Profits, and Lasting Value*. Harvard Business School Press: Boston, **1996.**

Samuel, M. O.; Chipunza, C. Employee Retention and Turnover: Using Motivational Variables as a Panacea. *Afr. J. Bus. Manage.* **2009,** *3*(8), 410–415.

Sempane, M.; Rieger, H.; Roodt, G. 'Job Satisfaction in Relation to Organisational Culture'. *South Afr. J. Ind. Psychol.* **2002,** *28*(2): 23–30.

Sousa-Poza, A.; Sousa-Poza, A. Well-being at Work: a Cross-National Analysis of the Levels and Determinants of Job Satisfaction. *J. Socio-Eco.* **2000,** *29*, 517–438.

Sutherland, M. Rethinking the retention of knowledge workers, Sept 2005, http://www. gibs.co.za/news-and-events/newsandevents_1/gibs-info-sharing/rethinking-the-retention-of-knowledge-workers.aspx_. (accessed March 15, 2012).

Tella, A.; Ayeni C. O.; Popoola S. O. Work Motivation, Job Satisfaction and Organizational Commitment of Library Personnel in Academic and Research Libraries in Oyo State, Nigeria. Library Philosophy and Practice, 2007.

Tzeng, H. The Influence of Nurses' Working Motivation and Job Satisfaction on Intention to Quit: An Empirical Investigation in Taiwan. *Int. J. Nurs. Studies* **2002,** *39*(8), 867–878.

Wegge, J.; Schmidt, K.; Parkes, C.; van Dick, R. "Taking a Sickie": Job Satisfaction and Job Involvement as Interactive Predictors of Absenteeism in a Public Organization. *J. Occup. Organ. Psychol.* **2007,** *80*(1), 77–89.

West, L. A.; Bogumil, W. A. Immigration and the Global IT Work Force. *Commun. ACM* **2001,** *44*(7), 34–39.

Yasin, M. Entrepreneurial Effectiveness and Achievement in Arab Culture. *J. Bus. Res.* **1996,** *35*(1), 69–77.

CHAPTER 5

FACTORS INFLUENCING CHANGING OF EMPLOYER BY HI-TECH PROFESSIONALS IN TECHNOLOGY-INTENSIVE INDUSTRIES

ŽIVA VEINGERL ČIČ[1,*], SAMO BOBEK[2], and
SIMONA ŠAROTAR ŽIŽEK[3]

[1]Faculty of Economics and Business, University of Maribor, Razlagova, Maribor, Slovenia, *E-mail: zivana.veingerl1@um.si

[2]Faculty of Economics and Business, University of Maribor, Razlagova, Maribor, Slovenia, E-mail: samo.bobek@um.si

[3]Faculty of Economics and Business, University of Maribor, Razlagova, Maribor, Slovenia, E-mail: simona.sarotar-zizek@um.si

CONTENTS

ABSTRACT

PURPOSE

Information technology (IT) is currently one of the most rewarding industries in the world. In the globalization and the boom in technology the organizations are operating in an unpredictable business environment where the workforce is more mobile, witnessing "war for talent" in the market place. Therefore, the employee retention of key employees is becoming one of the strategic HR goals. If employees are not satisfied and their company does not offer them opportunities for growth and development, their intention to quit will increase. If the organization doesn't recognize the signs of unsatisfied employees and provide them a reason to stay, they will leave the company. This chapter seeks to identify the relationship between job satisfaction and industry at employees in four countries (Austria, Germany, Slovenia, and Spain).

METHODOLOGY/APPROACH

The exploratory study is based on an online four-country survey, of 1207 employees who fulfilled the two criteria; first criterion was having a job at least 10 working hours per week and the second criterion was having colleagues at work. Among the respondents 294 were from Austria, 283 from Germany, 369 from Slovenia, and 261 from Spain.

FINDINGS

The statistical results of the study revealed that there aren't statistically significant differences regarding employee's intention to quit of employed IT and other activities.

ORIGINALITY/VALUE

This is the first study to analyses intention to quit in four countries perspective.

5.1   INTRODUCTION

Information technology (IT) is currently one of the most rewarding industries in the world. In globalization and the boom in technology the organizations are operating in an unpredictable business environment where the workforce is more mobile, witnessing "war for talent" in the market place. Therefore, employee retention of key employees is becoming one of the strategic human resources (HR) goals.

The Slovenian IT industry is one of the most penetrating and rapidly developing areas in Slovenia. For merit its progress can be attributed to HR and well-developed IT infrastructure, which is the result of early national realization of such industries as priority developments. The IT industry is a rapidly growing industry and the employee skills of hi-tech professionals are specific. If companies want to remain competitive in this area, they need to invest in their employees. This does not mean only to focus on getting experts from the IT fields, but it is necessary to provide activities which are focused on maintaining the hi-tech professionals. Key and prospective employees from the IT industry, where demand exceeded the offer of employment, need the reason to stay with the current company and not to look for another job opportunity. This, in other words, means that the company must recognize the potential of these employees and develop them, and at the same time, enable the individuals to realize their ambitions. If the company does not practice this, the key hi-tech employees will seek a new job opportunities elsewhere. Therefore, it is the intention of this paper to give suggestions for employers to reduce the intention to leave of hi-tech professionals. On the other hand, we have observed the connection of the intention to quit, and employee satisfaction in four countries (Austria, Germany, Slovenia and Spain). In doing so, we have focused on the characteristics and specificities of hi-tech professionals in four countries included in the study.

5.2   CONCEPTUAL BACKGROUND

The competition for the technically skilled individuals that are necessary to fill knowledge-based jobs has increased the importance of their recruitment due to sustained economic growth (Collins and Stevens, 2002). Moreover, census data indicates that demographic trends such as a smaller supply of younger workers and retirements among baby boomers will make it difficult to fill openings for the next decade, particularly those requiring technical and engineering skills (Dohm, 2000).

Managers will be in the future challenged by an increasingly mobile and transient workforce. There are four generations at the workplace, working together with different norms, believes, and motivation aspect. Especially young hi-tech professionals, generation Y and also X, and the coming generation Z will increasingly demand greater flexibility in their work and personal lives to work optimally. This is an important awareness, because these younger professionals will look elsewhere for work if they won't be able to satisfy their needs. Therefore, this is another important aspect to develop opportunities within the company.

Labor market of hi-tech professionals is quite specific. Employers should be aware that, according to survey of "MojeDelo" job search portal, less than half (48%) of respondents have obtained their last job without the active search (Zaletel, 2005). The employer has either directly taken a contact with them or the company has recommended an acquaintance. Companies, therefore, need to be aware that it is necessary to carry out specific activities in the field of HR management in order to maintain talented employees in the field of IT. The labor market is also interesting, because 37% of IT experts have "nontechnical" educational direction. The above mentioned study found that companies are not excessively strict in the selection conditions for IT staff (e.g., education), until they believe that the potential new employees can master the necessary technical knowledge. As a result, competition among organizations to obtain the highest quality professional IT staff will become even fiercer. Hi-tech professionals are workers with specific characteristics and therefore should be treated with special attention, both in attracting them to the company, as well as their motivation for work and cooperation.

These professionals are aware of their deficiency, so the companies are bargaining with them for wages and other benefits for their long-term stay; consequently, these employees often set the bar high. If the bargaining with them is unsuccessful, this can result in less loyalty to the company and also in higher turnover, especially if a competitive company is offering them better pay or better working conditions. There are some peculiarities of hi-tech professionals, such as that they have difficulties in tolerating authority, and consequently it is difficult to communicate with them. Gender plays no role in compensation for technology professionals.

That statement may seem shocking, however according to the analysis of Dice's annual salary survey (2016) data of more than 16,000 tech professionals, when comparing equal education levels, years of technical experience and job title, no gap exists. This study has shown that men and women in technology share the same concerns, receive bonuses, and are satisfied with their compensation. Although there is a difference: employers are offering to women more options like flexible work hours or the ability to telecommute versus men who are more offered compensation as a motivator. Due to the large demand for highly skilled hi-tech professionals and efforts required to keep those in the company, the costs of finding, hiring, and training these employees can deeply impact business' bottom line.

According to Dice's analysis of the latest turnover data from the U.S. Bureau of Labor Statistics, voluntary quits in Professional and Business services continue to trend upwards (2016). Total quits in the category in the first three quarters of 2015 averaged 508,200, the highest since 2002 and a 9% increase compared to the same period in 2014 (Dice, 2016).

Higher turnover means business and hi-tech professionals may have the leverage to leave their current jobs for a better position or higher pay and are more open to looking for a new job.

Understanding the priorities of hi-tech professionals, whether that can be greater work–life balance, flexible working options, or computer incentives will not only allow you to retain talent, but it will lead to a more engaged, productive, and ultimately profitable technology team.

5.3   METHODOLOGICAL ISSUES

We have developed following research question: Are there any differences in the intention to quit at present, regarding industry in four countries, Germany/Austria, Slovenia, and Spain?

The following are the specific issues which have been identified during this research study. Therefore, our main hypothesis is:

H1: Differences in the intention to quit regarding industry in four countries, Germany/Austria, Slovenia, and Spain are statistically significant.

Objectives of the study are depicted below:

- To identify the main factors of intention to quit of IT professionals.
- To identify profile of IT professionals.

- To identify the organization tools to win war for talent among IT professionals.

5.4 PROFILE OF IT EMPLOYEE

To identify key personality traits of IT professionals is very important, because they differentiate them from those in other occupations and can be used for a variety of purposes, such as the assessment of IT job candidates, for recruitment, selection, placement, training needs identification, career planning, counseling, and ongoing management.

Holland's (1996) theoretical framework characterizes IT professionals as having mainly investigative, realistic, and conventional interests as compared to other occupations, IT professionals are more: tough-minded and analytic, more open to new experiences and learning, emotionally resilient, customer-oriented, and intrinsically motivated. They are also less, conscientious, concerned with image management, and less visionary in their thinking style. Lounsbury et al. found that assertiveness, customer service orientation, optimism, and work drive of IT professionals are positively related to career satisfaction for individuals in a wide range of occupational fields, above and beyond the variance accounted for by the big five personality traits (2003). According to Myers & McCaulley IT personnel is populated predominantly with introverts, because of the independent tasks required and the fact that most IT jobs involve extended solitary tasks (1985). On the other hand, the survey of IT market in Slovenia in 2013 made by the company MojeDelo identified the characteristics that make up the so-called IT expert profile. These characteristics are as follows:

- Self-confident (reason: they are aware that there is still constant demand for IT experts in the market).
- Unfaithful (91% of the IT experts intend to leave their current job. The average length an IT expert stay on the same job is 2.2 years)
- Well educated (i.e., must educate)—Hi-tech professionals mostly educate themselves, so they are self-motivated to monitor and educate themselves about the latest trends of development tools and innovations in the field of IT. Most (75%) are dissatisfied, because companies do not invest enough in their education.

- They have problems transferring their authority (the biggest problem recruiters have with hi-tech professionals (and vice versa) is the communication of technical staff with other employees).
- Pay and rewards are their best motivators (67% of responses).

The determinants of employee turnover have great relevance to the employees who are thinking of quitting, as well as for the leaders who are faced with the lack of employee continuity. The high costs are usually involved in the induction and training of new personnel and the issue of organizational productivity (Firth et al., 2004; Appollis, 2010).

Hi-tech professionals exhibit characteristics that differ from those in other professions; they are more ambitious, logical, and conservative. They possess unique attitudes, interests, sense of identity, and work conscious-ness (Armstrong et al., 2007).

5.5   INTENTION TO QUIT

Purani and Sahadev (2007) define intention to quit as an employee plan for intention to leave the current job and as looking onwards to find another job in the near future. It can be seen as a worker's intention to leave the present organization (Cho et al., 2009), or as an "individual's own estimated prob-ability" (subjective) that s/he is permanently leaving the organization at some point in the near future (Vandenberg and Nelson 1999) and it can be seen as the strength of an individual's view that s/he does not want to stay with his/her current employer (Boshoff et al., 2002). With the concept "intention to quit" has to be mentioned that Sommer and Haug recognized the intention to leave as the immediate predictor to actual behavior of an employee (2010). Although having intention to leave the company will not necessarily lead to actual leaving, but the intention to leave itself will affect organizational performance because it contributed to work ineffi-ciencies, disengagement, and absenteeism (Kivimaki et al., 2007).

In the IT sector, the issues of intention to leave have not just occur recently, they began since early 2000s and although the salaries and benefits associated with IT career continue to increase, the number of employees leaving their current companies did not decrease (Rouse, 2001). A meta-analyses (Rouse and Corbitt 2007) identified that the job satisfaction, job performance, role conflict, pay, promotion, and perceived job characteristics are the important factors for quitting intentions of IT professionals.

IT workers are a unique class of workers and have been viewed as such since the 1970s (Armstrong et al., 2007). Nearly every organization needs IT workers (Ranii, 2012) and the skills clusters of hi-tech professionals enable opportunities for their frequent job changes (Gabe and Abel, 2011). Developing new skills adds to the strategic value of IT workers, earning them higher salaries and making them less susceptible to the effects of economic recession (Janz and Nichols, 2010; Morrissey, 2011). But the organization can influence the intention and decision to quit or to stay with three main factors:

- A liberated and empowered culture
- Equity and enablement for high performance
- As well as an effective and interactive communication channel (Gaylard et al., 2005)

Nevertheless, we can add other factors in the organizations that can influence the IT worker's decision to leave or stay. These factors are: (1) job satisfaction, (2) financial reward, (3) employability and personal growth, (4) the job itself, (5) relationship with the boss, and the (5) organizational culture and environment (ibid). On the other hand, Armstrong et al. (2007) stated that other factors can influence the decision to leave, such as barriers to promotion, turnover, managing family, work stress, work schedule flexibility, and job qualities.

5.5.1 THEORETICAL MODELS: INTENTION TO QUIT

In this chapter, we focus on three models of intention to quit that describe the complex nature of intention to quit and are the foundation to develop a theoretical model of intention to quit. These models are:

1. Oehley's talent management model (Oehley, 2007),
2. Dhladhla's model of the influence of leader behavior, psychological empowerment, job satisfaction, and organizational commitment on turnover intention, (Dhladhla's, 2011), and
3. Berry's model of employee engagement, compensation fairness, job satisfaction, and turnover intent (Berry, 2010).

5.5.1.1   OEHLEY'S TALENT MANAGEMENT MODEL

The focus of this model is the development and evaluation of a partial talent management competency model. The key objectives of this model were to identify talent management competencies required by line managers for successful implementation of an organization's talent management strategy. The further objective was to determine how line managers' talent management competencies affect subordinates' intention to remain with the organization. Organizational job satisfaction has a significant effect on intention to quit, concluded Oehley (2007). Therefore, this model is an important part of the talent management competencies, because it includes attracting, developing, and retaining talented employees, which is an essential responsibility of a leader.

5.5.1.2   DHLADHLA'S MODEL OF THE INFLUENCE OF LEADER BEHAVIOR, PSYCHOLOGICAL EMPOWERMENT, JOB SATISFACTION AND ORGANIZATIONAL COMMITMENT ON TURNOVER INTENTION

This model determines the causal linkages between leader behavior, psychological empowerment, job satisfaction, and organizational commitment and their influence on turnover intention (Dhladhla, 2011). Dhladhla advised that the study had the potential for expansion through the inclusion of further variables that are likely to influence turnover intention among employees (2011).

5.5.1.3   BERRY'S MODEL OF EMPLOYEE ENGAGEMENT, COMPENSATION FAIRNESS, JOB SATISFACTION, AND TURNOVER INTENT

Berry's model assessed the effects of age and the mediating effects of job satisfaction on the relationship between employee engagement, compensation fairness, and the outcome variable turnover intent. Berry's indicated that both employee engagement and compensation fairness are inversely related to turnover intent and that job satisfaction did not mediate the relationship between employee engagement and intention to quit or that between compensation fairness and intention to quit (2010).All illustrated models focus on various variables, with the common purpose of better understanding the relationship between the independent variables and the investigated dependent variable, intention to quit.

5.6   THE RELATIONSHIP BETWEEN JOB SATISFACTION AND INTENTION TO QUIT

Job satisfaction can influence a variety of important attitudes, intentions, and behaviors. One of the most comprehensive and widely used measures for job satisfaction was presented by Hunt et al. (1985) and Purani and Sahadev (2007). They used a job satisfaction as predictor variable and examined its impact on intention to leave. Purani and Sahadev (2007) found that employees with long stay at workplace had higher level of job satisfaction and would not be inclined to quit. Job satisfaction can be characterized with six major facets (Gemadde and Buddhika, 2013):

1.   Satisfaction with supervisor (determines the level of job satisfaction on the basis of employees' perception on how much they are satisfied with the guidelines provided to them by their supervisors).
2.   Satisfaction with variety (having a variety of tasks such as challenging not routine, the opportunities available for them to grow in the organization, this dimension also measures the employee perception of job satisfaction through the level of perceived freedom in job).
3.   Satisfaction with closure.
4.   Satisfaction with compensation (compensation is one of the most extrinsic indicators of job satisfaction because it determines the level of job satisfaction of employees by knowing how much they are satisfied with the pay or compensation or any other security their jobs have provided to them).
5.   Satisfaction with coworkers (how an employee perceives his/her job accomplishment by the support or the presence of his/her coworker's attitude and behavior such as selfishness, friendliness, or supportiveness).
6.   Satisfaction with management and HR policies.

Several theories tried to explain the link between job satisfaction and turnover intention. The theory of turnover sees turnover intention as an evaluation of the current and expected working conditions (Mobley, 1982).

Social identity approach and self-categorization theory explain the relation between job satisfaction and turnover intention: Satisfied coworkers identify with their organization and hence the coworkers' own future is determined by the organization's future more than before (Van Dick et al., 2004).

Zurn et al. pointed out that the shortages of workforce in a specific sector or organization can be a symptom of low job satisfaction, poor management, and lack of organizational support (2005).

There is still an unresolved debate as to whether job satisfaction has a direct effect on intention to quit. It is very likely that high levels of dissatisfaction could influence employees to consider alternative job options. Whether an employee will really leave an organization in such a case, or has an intention be apparent, is in most cases determined by alternative opportunities in the labor market (Spector, 1997). Job dissatisfaction not only has an effect on the organization, as it increases the intention to quit of employees, but it also reduces the contribution of the employee to the organization, which has a direct effect on the organization's success (Lok and Crawford, 2003).

5.7   QUALITATIVE RESEARCH

5.7.1   SAMPLE AND PROCEDURE

Austrian and Slovene workers were invited to participate in an online study. Data were collected within the culture leadership project funded by the Styria federal state within the "grenz-frei" (= 'no border') framework. The invitation to the online study was sent out in cooperation with a well-known German market research company. Participants who did not fulfill the requirements of (I) having a job with at least 10 working hours per week and (II) having colleagues at work, were selected at the beginning of the study and weren't able to participate in the online study further on.

All in all, 2407 persons were invited to participate in the online study. 1207 persons fulfilled the criteria (having a job with at least 10 working hours per week and having colleagues at work) and started the questionnaire. 294 participants came from Austria, 261 were from Spain, 369 worked in Slovenia, and 283 came from Germany. The online study started at the 9th of July and was finished at the 19th of July 2013.

TABLE 5.1   Respondents Regarding Country. (*Source*: Authors)

Country		Frequency	Percent	Valid percent	Cumulative percent
Valid	Austria	258	24.5	24.5	24.5
	Spain	247	23.4	23.4	47.9
	Slovenia	292	27.7	27.7	75.5
	Germany	258	24.5	24.5	100.0
	Total	1055	100.0	100.0	

The largest proportion of respondents (Table 5.1), almost 28% were from Slovenia, followed by 24.5% of Austria and Germany respondents and 23.4% were respondents from Spain. Table 5.2 shows, that, among all surveyed more than 54% were female respondents and slightly less than 46% were male respondents.

TABLE 5.2   Respondents Regarding Sex. (*Source*: Authors)

Sex		Frequency	Percent	Valid percent	Cumulative percent
Valid	Male	482	45.7	45.7	45.7
	Female	573	54.3	54.3	100.0
	Total	1055	100.0	100.0	

TABLE 5.3   Respondents Regarding Age. (*Source*: Authors)

Age in years		Frequency	Percent	Valid percent	Cumulative percent
Valid	Until 20	14	1.3	1.3	1.3
	21–25	58	5.5	5.5	6.8
	26–30	139	13.2	13.2	20.0
	31–35	178	16.9	16.9	36.9
	36–40	165	15.6	15.6	52.5
	41–45	142	13.5	13.5	66.0
	46–50	125	11.8	11.8	77.8
	51–55	112	10.6	10.6	88.4
	56–60	81	7.7	7.7	96.1
	61–65	38	3.6	3.6	99.7
	66 and older	3	0.3	0.3	100.0
	Total	1055	100.0	100.0	

Most of the respondents were in the age group from 31 to 45 years of age (46% of all respondents).

TABLE 5.4 Respondents Regarding Education. (*Source*: Authors)

		Highest education			
		Frequency	Percent	Valid percent	Cumulative percent
Valid	Compulsory education	20	1.9	1.9	1.9
	Apprenticeship	281	26.6	26.6	28.5
	Finished grammar/ secondary school/ college	250	23.7	23.7	52.2
	University	504	47.8	47.8	100.0
	Total	1055	100.0	100.0	

University degree had 47.8% of all respondents, followed by apprenticeship (26.6%).

5.7.2 RESULTS

We have used quantitative methods (descriptive statistics) to research intention to quit at employees in four countries. Presented results of descriptive statistics are in Table 5.5.

Based on Table 5.5, it is indicated that the statement "All in all satisfaction with my work…" has the highest average value of 2.51, while the assertion "I would prefer working in a different business" achieves the highest average value of the credit and insurance. In the sector craftsmanship and building/construction, the highest average value has three arguments as follows:

1. "In the last year, only a few people were dismissed and my company"
2. "I thought of changing to a passing department within my company" and
3. "I am afraid to lose my current job."

Health sector reached the highest average value in the argument, "The thought of looking for a new job already entered my mind."

If we look at the average value of individual items according to the industry, then we see that IT achieves above average value for all items of construct intention to quit.

TABLE 5.5 Descriptive Statistics of Items in Construct Intention to Quit Regarding Industry. (*Source*: Authors)

		N	Mean	Std. devia- tion	Std. error	95% Confidence interval for mean	
						Lower bound	Upper bound
All in all, satis- faction with my work…	Public sector	100	2.51	1.087	0.109	2.29	2.73
	Industry	115	2.39	0.915	0.085	2.22	2.56
	Health care sector	65	2.32	1.002	0.124	2.07	2.57
	Commerce/trades	68	2.57	1.055	0.128	2.32	2.83
	Education	72	2.19	0.762	0.090	2.02	2.37
	Gastronomy and tourism	30	2.43	1.135	0.207	2.01	2.86
	Credit and insurance	39	2.23	0.872	0.140	1.95	2.51
	Craftsmanship and building/ construction	36	2.44	1.054	0.176	2.09	2.80
	IT	89	2.62	1.017	0.108	2.40	2.83
	Other services	67	2.54	1.064	0.130	2.28	2.80
	Logistics/ transport	40	2.45	1.061	0.168	2.11	2.79
	Other	185	2.45	1.068	0.078	2.29	2.60
	999	38	2.68	0.962	0.156	2.37	3.00
	Total	944	2.45	1.012	0.033	2.39	2.52

TABLE 5.5 *(Continued)*

		N	Mean	Std. devia-tion	Std. error	95% Confidence interval for mean	
						Lower bound	Upper bound
I thought of changing to a different depart-ment within my company.	Public sector	100	2.75	1.077	0.108	2.54	2.96
	Industry	115	2.78	1.016	0.095	2.60	2.97
	Health care sector	65	2.98	1.139	0.141	2.70	3.27
	Commerce/trades	68	2.81	1.026	0.124	2.56	3.06
	Education	72	3.06	1.033	0.122	2.81	3.30
	Gastronomy and tourism	30	3.10	1.155	0.211	2.67	3.53
	Credit and insurance	39	2.85	0.988	0.158	2.53	3.17
	Craftsmanship and building/ construction	36	3.33	0.828	0.138	3.05	3.61
	IT	89	2.89	0.959	0.102	2.69	3.09
	Other services	67	3.01	1.007	0.123	2.77	3.26
	Logistics/ transport	40	2.90	0.928	0.147	2.60	3.20
	Other	185	3.03	1.040	0.076	2.88	3.18
	999	38	2.45	1.058	0.172	2.10	2.80
	Total	944	2.92	1.035	0.034	2.85	2.98

TABLE 5.5 *(Continued)*

		N	Mean	Std. devia-tion	Std. error	95% Confidence interval for mean	
						Lower bound	Upper bound
The thought of looking for a new job already entered my mind.	Public sector	100	2.54	1.114	0.111	2.32	2.76
	Industry	115	2.63	1.055	0.098	2.43	2.82
	Health care sector	65	2.57	1.145	0.142	2.29	2.85
	Commerce/trades	68	2.56	1.238	0.150	2.26	2.86
	Education	72	2.67	1.113	0.131	2.41	2.93
	Gastronomy and tourism	30	2.23	1.135	0.207	1.81	2.66
	Credit and insurance	39	2.82	1.189	0.190	2.44	3.21
	Craftsmanship and building/construction	36	2.75	1.228	0.205	2.33	3.17
	IT	89	2.56	1.087	0.115	2.33	2.79
	Other services	67	2.30	1.142	0.139	2.02	2.58
	Logistics/transport	40	2.65	1.145	0.181	2.28	3.02
	Other	185	2.63	1.169	0.086	2.46	2.80
	999	38	2.34	0.994	0.161	2.02	2.67
	Total	944	2.57	1.134	0.037	2.50	2.64

TABLE 5.5   *(Continued)*

		N	Mean	Std. devia-tion	Std. error	95% Confidence interval for mean	
						Lower bound	Upper bound
I would prefer working in a different business.	Public sector	100	2.67	1.120	0.112	2.45	2.89
	Industry	115	2.97	1.072	0.100	2.78	3.17
	Health care sector	65	3.18	0.967	0.120	2.95	3.42
	Commerce/trades	68	2.74	1.115	0.135	2.47	3.01
	Education	72	3.07	0.969	0.114	2.84	3.30
	Gastronomy and tourism	30	2.47	1.224	0.224	2.01	2.92
	Credit and insurance	39	3.15	0.988	0.158	2.83	3.47
	Craftsmanship and building/ construction	36	3.14	1.046	0.174	2.78	3.49
	IT	89	3.03	0.947	0.100	2.83	3.23
	Other services	67	2.66	1.109	0.135	2.39	2.93
	Logistics/ transport	40	2.83	1.083	0.171	2.48	3.17
	Other	185	2.79	1.084	0.080	2.64	2.95
	999	38	2.50	0.923	0.150	2.20	2.80
	Total	944	2.87	1.068	0.035	2.80	2.93

TABLE 5.5   *(Continued)*

		N	Mean	Std. devia-tion	Std. error	95% Confidence interval for mean	
						Lower bound	Upper bound
I have already been looking for a new job.	Public sector	100	2.90	1.142	0.114	2.67	3.13
	Industry	115	3.00	1.060	0.099	2.80	3.20
	Health care sector	65	2.80	1.214	0.151	2.50	3.10
	Commerce/trades	68	2.84	1.288	0.156	2.53	3.15
	Education	72	3.07	1.105	0.130	2.81	3.33
	Gastronomy and tourism	30	2.77	1.194	0.218	2.32	3.21
	Credit and insurance	39	3.13	1.151	0.184	2.76	3.50
	Craftsmanship and building/construction	36	2.86	1.222	0.204	2.45	3.27
	IT	89	3.09	1.104	0.117	2.86	3.32
	Other services	67	2.72	1.204	0.147	2.42	3.01
	Logistics/transport	40	3.15	1.051	0.166	2.81	3.49
	Other	185	2.91	1.167	0.086	2.74	3.08
	999	38	2.63	0.998	0.162	2.30	2.96
	Total	944	2.92	1.148	0.037	2.85	3.00

TABLE 5.5 *(Continued)*

		N	Mean	Std. devia-tion	Std. error	95% Confidence interval for mean	
						Lower bound	Upper bound
I would like to change my job within the next half year.	Public sector	100	3.04	1.044	0.104	2.83	3.25
	Industry	115	3.16	0.961	0.090	2.98	3.33
	Health care sector	65	3.00	1.132	0.140	2.72	3.28
	Commerce/trades	68	2.87	1.183	0.143	2.58	3.15
	Education	72	3.22	0.938	0.111	3.00	3.44
	Gastronomy and tourism	30	2.80	1.095	0.200	2.39	3.21
	Credit and insurance	39	3.23	1.012	0.162	2.90	3.56
	Craftsmanship and building/ construction	36	3.06	1.218	0.203	2.64	3.47
	IT	89	3.13	1.025	0.109	2.92	3.35
	Other services	67	2.73	1.123	0.137	2.46	3.01
	Logistics/ transport	40	3.00	1.086	0.172	2.65	3.35
	Other	185	2.97	1.151	0.085	2.80	3.13
	999	38	2.58	1.004	0.163	2.25	2.91
	Total	944	3.01	1.082	0.035	2.94	3.08

TABLE 5.5   *(Continued)*

		N	Mean	Std. devia-tion	Std. error	95% Confidence interval for mean	
						Lower bound	Upper bound
I am afraid to lose my current job.	Public sector	100	2.97	1.000	0.100	2.77	3.17
	Industry	115	3.00	1.009	0.094	2.81	3.19
	Health care sector	65	3.14	0.882	0.109	2.92	3.36
	Commerce/trades	68	2.66	1.060	0.128	2.41	2.92
	Education	72	2.97	1.021	0.120	2.73	3.21
	Gastronomy and tourism	30	3.30	0.837	0.153	2.99	3.61
	Credit and insurance	39	2.87	1.080	0.173	2.52	3.22
	Craftsmanship and building/construction	36	2.86	1.046	0.174	2.51	3.22
	IT	89	2.83	1.014	0.107	2.62	3.05
	Other services	67	2.84	1.067	0.130	2.58	3.10
	Logistics/transport	40	2.85	1.189	0.188	2.47	3.23
	Other	185	2.75	1.065	0.078	2.60	2.91
	999	38	2.53	1.006	0.163	2.20	2.86
	Total	944	2.88	1.034	0.034	2.81	2.94

TABLE 5.5   *(Continued)*

		N	Mean	Std. devia-tion	Std. error	95% Confidence interval for mean	
						Lower bound	Upper bound
In the last year, only a few people were dismissed in my company.	Public sector	100	2.06	1.127	0.113	1.84	2.28
	Industry	115	1.85	1.053	0.098	1.66	2.05
	Health care sector	65	1.83	1.069	0.133	1.57	2.10
	Commerce/trades	68	2.18	1.145	0.139	1.90	2.45
	Education	72	1.96	1.013	0.119	1.72	2.20
	Gastronomy and tourism	30	1.87	1.008	0.184	1.49	2.24
	Credit and insurance	39	1.92	1.010	0.162	1.60	2.25
	Craftsmanship and building/ construction	36	2.42	1.204	0.201	2.01	2.82
	IT	89	2.19	1.117	0.118	1.96	2.43
	Other services	67	2.06	1.013	0.124	1.81	2.31
	Logistics/ transport	40	2.23	1.097	0.174	1.87	2.58
	Other	185	2.12	1.131	0.083	1.95	2.28
	999	38	1.97	0.944	0.153	1.66	2.28
	Total	944	2.05	1.089	0.035	1.98	2.12

Due to research differences in intention to quit regarding industry, it is important to perform ANOVA. Results of this test are presented in Table 5.6.

TABLE 5.6   Presents the Results of One-Way Analysis of Variance (ANOVA). (*Source*: Authors)

		The sum of the squares	df	Average squares	F	p
		ANOVA				
All in all satisfaction with my work…	Between groups	14.532	12	1.211	1.185	0.289
	Within groups	951.226	931	1.022		
	Total	965.758	943			
I thought of changing to a different department within my company.	Between groups	26.130	12	2.178	2.062	0.017
	Within groups	983.090	931	1.056		
	Total	1009.220	943			
The thought of looking for a new job already entered my mind.	Between groups	15.919	12	1.327	1.032	0.417
	Within groups	1197.183	931	1.286		
	Total	1213.102	943			
I would prefer working in a different business.	Between groups	38.156	12	3.180	2.855	0.001
	Within groups	1037.027	931	1.114		
	Total	1075.182	943			
I've already been looking for a new job.	Between groups	16.904	12	1.409	1.069	0.383
	Within groups	1226.756	931	1.318		
	Total	1243.660	943			
I would like to change my job within the next half year.	Between groups	24.459	12	2.038	1.758	0.051
	Within groups	1079.503	931	1.160		
	Total	1103.962	943			
I am afraid to lose my current job.	Between groups	24.151	12	2.013	1.905	0.030
	Within groups	983.595	931	1.056		
	Total	1007.746	943			

TABLE 5.6   *(Continued)*

		ANOVA				
		The sum of the squares	df	Average squares	F	p
In the last year, only a few people were dismissed in my company.	Between groups	19.903	12	1.659	1.407	0.157
	Within groups	1097.855	931	1.179		
	Total	1117.758	943			

In Table 5.6, we can see the calculation of statistical significance of p. The statements exceeding 0.05 are:

- All-in-all satisfaction with my work
- The thought of looking for a new job already entered my mind.
- Already I've been looking for a new job.
- In the last year, only a few people were dismissed in my company.

By the statement where the calculated statistical significance is $p < 0.05$, we can talk about statistically significant differences with consideration to sector:

- I thought of changing department within my company.
- I would prefer working in a different business.
- I am afraid to lose my current job.

For the statement "I would like to change my job within the next half year" p is $p < 0.051$.

Since we were interested in which groups data vary significantly, we performed an additional so-called post hoc analysis, which was based on the Tukey HSD method. These tests have shown that there are no statistically significant differences in terms of activity in any of the statements in Table 5.7.

TABLE 5.7 Results of Tukey HSD Method for "I Thought of Changing to a Different Department within My Company." (*Source*: Authors)

Industrial sector categorized		N	Subset for alpha = 0.05	
			1	2
Tukey HSD[a,b]	999	38	2.45	
	Public sector	100	2.75	2.75
	Industry	115	2.78	2.78
	Commerce/trades	68	2.81	2.81
	Credit and insurance	39	2.85	2.85
	IT	89	2.89	2.89
	Logistics/transport	40	2.90	2.90
	Health care sector	65	2.98	2.98
	Other services	67	3.01	3.01
	Other	185	3.03	3.03
	Education	72	3.06	3.06
	Gastronomy and tourism	30		3.10
	Craftsmanship and Building/construction	36		3.33
	Sig.		0.088	0.124

Means for groups in homogeneous subsets are displayed. [a]Uses harmonic mean sample size = 55.960. [b]The group sizes are unequal. The harmonic mean of the group sizes is used. Type I error levels are not guaranteed.

TABLE 5.8 Results of Tukey HSD Method for Item "I Would Prefer Working in a Different Business." (*Source*: Authors)

Industrial sector categorized		N	Subset for alpha = 0.05		
			1	2	3
Tukey HSD[a,b]	Gastronomy and tourism	30	2.47		
	999	38	2.50	2.50	
	Other services	67	2.66	2.66	2.66
	Public sector	100	2.67	2.67	2.67
	Commerce/trades	68	2.74	2.74	2.74
	Other	185	2.79	2.79	2.79
	Logistics/transport	40	2.83	2.83	2.83
	Industry	115	2.97	2.97	2.97
	IT	89	3.03	3.03	3.03
	Education	72	3.07	3.07	3.07
	Craftsmanship and building/construction	36		3.14	3.14
	Credit and insurance	39		3.15	3.15
	Health care sector	65			3.18
	Sig.		0.118	0.057	0.284

Means for groups in homogeneous subsets are displayed. [a]Uses harmonic mean sample size = 55.960. [b]The group sizes are unequal. The harmonic mean of the group sizes is used. Type I error levels are not guaranteed.

TABLE 5.9   Results of Tukey HSD Method for "I am Afraid to Lose My Current Job". (*Source*: Authors)

Industrial sector categorized		N	Subset for alpha = 0.05	
			1	2
Tukey HSD[a,b]	999	38	2.53	
	Commerce/trades	68	2.66	2.66
	Other	185	2.75	2.75
	IT	89	2.83	2.83
	Other services	67	2.84	2.84
	Logistics/transport	40	2.85	2.85
	Craftsmanship and building/ construction	36	2.86	2.86
	Credit and insurance	39	2.87	2.87
	Public sector	100	2.97	2.97
	Education	72	2.97	2.97
	Industry	115	3.00	3.00
	Health care sector	65	3.14	3.14
	Gastronomy and tourism	30		3.30
	Sig.		0.083	0.056

Means for groups in homogeneous subsets are displayed. [a]Uses harmonic mean sample size = 55.960. [b]The group sizes are unequal. The harmonic mean of the group sizes is used. Type I error levels are not guaranteed.

5.8   CONCLUSION

Findings of this research indicated that hi-tech professionals' intention to leave an organization in four researched countries is not showing statistically significant differences regarding sector in which employees work. Intention to quit is presented also in IT (all items have high average value).

Although there were no statistical differences between four countries in hi-tech professionals' intention to quit, we can overall suggest some recommendations to reduce the hi-tech employees' intention to leave their current organizations. There are also coordinated activities of management and HRD required for attracting and retaining employees. If hi-tech professionals are not given opportunities to realize their potential and ambitions, they will find work elsewhere because they are the personnel by which, despite the general economic crisis, demand still exceeds supply. Therefore, firms should act in two directions: to carry out HR activities to increase employee satisfaction and commitment that they remain in the company; on the other hand, they need managers with the appropriate transformational leadership competencies to further affect the well-being and satisfaction of these employees. It is important that the management identify the job satisfaction factors of the employees and take necessary actions to reduce the dissatisfaction of the employees. Profound identification of job satisfaction factors is very important since these factors may affect each employee in different ways. Some factors can even be resolved without any cost.

Identification of key talents in organization can help to promote competitive advantage and, therefore, enterprises should let organizations choose employees with personality characteristics such as the ability to endure high pressure; possess creativity and provide new constructive ideas and know-how; and actively overcome external obstructions (Brissett and Nowick, 1976).

5.9 FURTHER RESEARCH

This study can be used as a starting point for deeper research of hi-tech professionals and improved human resource tools to retain these employees and succeed in war for talents. There are many other perspectives that can be an interesting basis to look deeper into and to research further. This study only emphasized potential determinants that may influence employees to leave a job without reference to their position. It could be interesting to study the difference between managerial or nonmanagerial intent to leave or those who work part-time or full time only. It could also be of interest to look at role of companies in retaining their employees and what is the connection between the job satisfaction and intention to leave. Some further research can be made to examine different views of an intention to leave a job of managers and non-managers perspectives.

KEYWORDS

- **intention quit**
- **job satisfaction**
- **IT industry**
- **Austria**
- **Germany**
- **Slovenia**
- **Spain**

REFERENCES

Appollis, V. P. *The Relation Between Intention to Quit, Psychological Capital and Job Satisfaction in the Tourism Industry in the Western Cape;* Magister commercii. University of the Western Cape, EMS Faculty: Bellville, 2010.

Armstrong, D. J.; Riemenschneider, C. K.; Allen, M. W.; Reid, M. F. Advancement, Voluntary Turnover and Women in IT: A Cognitive Study of Work—Family Conflict. *Inf. Manage.* **2007,** *44*(2), 142–153.

Berry, M. L. *Predicting Turnover Intent: Examining the Effects of Employee Engagement, Compensation Fairness, Job Satisfaction, and Age;* Doctoral Dissertation. University of Tennessee: Knoxville, 2010.

Boshoff, A. B.; Van Wyk, R.; Hoole, C.; Owen, J. H. The Prediction of Intention to Quit by Means of Biographic Variables, Work Commitment, Role Strain and Psychological Climate. *Manage. Dyn.* **2002,** *11*(4), 14–28.

Brissett, M.; Nowicki, S. Internal vs. External of Reinforcement and Reaction to Frustration. *J. Abnorm. Soc. Psychol.* **1976,** *25,* 35–39.

Cho, S.; Johanson, M. M.; Guchait, P. Employees Intent to Leave: A Comparison of Determinants of Intent to Leave Versus Intent to Stay. *Int. J. Hospitality Manage.* **2009,** *28,* 374–381.

Collins, J. C.; Stevens, C. K. The Relationship Between Early Recruitment-Related Activities and the Application Decision of New Labor-Market Entrants: A Brand Equity Approach to Recruitment. *J. Appl. Psychol.* **2002,** *87*(6), 1121–1133.

Coomber, B.; Barriball, K. L. Impact of Job Satisfaction Components on Intent to Leave and Turnover for Hospital-Based Nurses: A Review of the Research Literature. *Int. J. Nurs. Stud.* **2007,** *44,* 297–314.

Dhladhla, T. J. The Influence of Leader Behaviour, Psychological Empowerment, Job Satisfaction, and Organizational Commitment on Turnover Intention. Unpublished master's thesis, Stellenbosch University, 2011.

Dice's annual salary survey, 2016, http://media.dice.com/report/march-2016-the-gender-factor-in-tech/. (accessed April 20, 2016).

Dohm, A. Gauging The Labor Force Effects of Retiring Baby-Boomers. *Mon. Labor Rev.* **2000,** *123*(7), 17–25.

Firth, L.; Mellor, D.; Moore, K.; Loquet, C. How Can Managers Reduce Employee Intention to Quit? *J. Managerial Psychol.* **2004,** *19*(2), 170–187.

Gabe, T.; Abel, J. Ragglomeration of Knowledge. *Urban Studies (Sage Publications, Ltd.)* **2011,** *48*(7), 1353–1371.

Gaylard, M.; Sutherland, M.; Viedge, C. The Factors Perceived to Influence the Retention of Information Technology Workers. *S. Afr. J. Bus. Manage.* **2005,** *36*(3), 87–97.

Gemadde, P.; Buddhika, K. D. M. Job Satisfaction and Intention to Leave of it Professionals in Sri Lanka 2013. *Asia Pac. J. Mark. Manage. Rev.* **2013,** *2*(9).

Holland, J. L. Exploring Careers with a Typology: What we have Learned and Some New Directions. *Am. Psychol.* **1996,** *51*(4), 397–406.

Hunt, S. D.; Chonko, L. B.; Wood, V. R. Organizational Commitment and Marketing. *J. Mark.* **1985,** *49*(1), 112–126.

Janz, B. D.; Nichols, E. L. Meeting the Demand for IT Workers: Can Career Choice be Managed? SIGMIS CPR'10: Proceedings of the 2010 ACM SIGMIS, 2010.

Kivimaki, M.; Vanhala, A.; Pentti, J.; Lansisalmi, H.; Virtanen, M.; Elovainio, M., et al. Team Climate, Intention to Leave and Turnover Among Hospital Employees: Prospective Cohort Study. *BMC Health Serv. Res.* **2007,** *7,* 170.

Lok, P.; Crawford, J. The effect of organizational culture and leadership style on job satisfaction and organizational commitment: A cross-national comparison. Journal of Management Development, 2003, 23, 321-338.

Lounsbury, J. W.; Loveland, J. M.; Sundstrom, E.; Gibson, L. W.; Drost, A. W.; Hamrick, F. An Investigation of Personality Traits in Relation to Career Satisfaction. *J. Career Assess.* **2003,** *11*(3), 287–307.

Mobley, W. CoWorker Turnover: Causes, Consequences, and Control. Addison-Wesley: Reading, MA, 1982.

Morrissey, J. Experts in Short Supply. *Trustee* **2011,** *64*(2), 13–16.

Myers, I. B.; McCaulley, M. H. *Manual: A Guide to the Development and Use of The Myers-Briggs Type Indicator;* Consulting Psychologists Press: Palo Alto, CA, 1985.

Oehley, A-M. *The Development and Evaluation of a Partial Talent Management Competency Model.* Unpublished Master's thesis, Stellenbosch University, 2007.

Purani, K.; Sahadev, S. The Moderating Role of Industrial Experience in the Job Satisfaction, Intention to Leave Relationship: An Empirical Study Among Salesmen in India. *J. Bus. Ind. Mark.* **2007,** *23*(7), 475–485.

Ranii, D. Some Job Candidates Can Still Call the Shots. The News and Observer: Raleigh, NC, Feb 26, 2012; pp 1–2.

Rouse, P. D. Voluntary Turnover Related to Information Technology Professionals: A Review of Rational and Instinctual models. *Int. J. Organ. Anal.* **2001,** *9*(3), 281–290.

Rouse, A. C.; Corbitt, B. J. Understanding information systems outsourcing success and risks through the lens of cognitive biases. Proceedings of the fifteenth European conference on information systems (ECIS), St Gallen, Switzerland, **2007,** June 7–9.

Spector, P. E. Job satisfaction: Application, assessment, causes, and consequences. Thousand Oaks, CA: Sage, **1997.**

Sommer, L.; Haug, M. Intention as Cognitive Antecedent to International Entrepreneurship—Understanding the Moderating Role of Knowledge and Experience. *Int. Entrepreneurship Manage. J.* **2010,** *7*(1), 111–142.

Van Dick, R.; Christ, O.; Stellmacher, J.; Wagner, U. J.; Ahlswede, O.; Grubba, C.; Hauptmeier, M.; Höhfeld, C.; Moltzen, K.; Tissington, P. Should I Stay or Should I Go? Explaining Turnover Intentions with Organizational Identification and Job Satisfaction. *Br. J. Manage.* **2004,** *15,* 351–360.

Vandenberg, R. J.; Nelson, J. B. Disaggregating the Motives Underlying Turnover Intentions: When do Intentions Predict Turnover Behaviour?. *Hum. Relat.* **1999,** *52,* 1313–1336.

Zaletel, A. *Raziskava IT trga v Sloveniji;* 2005. http://www.revija.mojedelo.com/hr/raziskava-it-trga-v-sloveniji-66.aspx (accessed April 18, 2016).

Zurn, P.; Dolea, C.; Stillwell, B. Nurse Retention and Recruitment: Developing a Motivated Workforce, In The Global Nursing Review Initiative. International Council of Nurses, 6, 2005.

CHAPTER 6

STRESS AS AN INEVITABLE PART OF WORK IN THE PRESENT AND FUTURE: STUDY OF OCCUPATIONAL STRESS AMONG NURSES, OFFICE CLERKS, AND TEACHERS

SHEIKH ABUL BARKAT[1,*] and GIRIJESH KUMAR YADAV[2]

[1]Regional Director, Maulana Azad National Urdu University, Mumbai Regional Centre, Plot No-60, Lane-G, Sector-8, Vashi, Navi Mumbai 400703, India, *E-mail: abulbarkat_2006@yahoo.co.in

[2]Technical Officer-A (Psychology), ICMR-National Institute of Occupational Health, Department of Health Research, Ministry of Health & Family Welfare, Govt. of India, Ahmedabad 380016, India, E-mail: girijeshkrydv@gmail.com

CONTENTS

ABSTRACT

The present chapter investigates the effect of type of profession on the occupational stress of working women. For this study, 270 working women have been taken from three different professions, that is, nursing, teaching, and office clerks from some eastern districts of Uttar Pradesh ranging between 30 and 45 years of age. The occupational stress scale (Singh and Srivastava, 1981) has been administered over the respondents to assess the level of their occupational stress. Analysis of variance (ANOVA) revealed the significant difference in the mean of working women belonging to nursing, office clerk, and teaching professions. Nursing women were found to be more stressed towards their job then followed by office clerks and teachers.

6.1 INTRODUCTION

Stress is possibly the most common problem of everyday life. The term "Stress" is so everywhere that it is used as a noun when we talk about being under "stress," as a verb when events are "stressing" us, and as an adjective when modern life has become "stressful."

Stress is a concept, which is familiar to both layman and professional alike. It may be understood that the stress is a state of mental or emotional strain or tension resulting from adverse or demanding circumstances. The concise Oxford Dictionary defines stress in five different ways. Three of them are of interest here. The first definition offered is that of a constraining or impelling force, and one example used is "under the stress of poverty." The second definition treats it as an effort or demands on energy, as in "subjected to great stress." The third definition offered talks of the force exerted on a body.

In recent years, research in medical and psychological fields has demonstrated that the widespread concern about stress at the workplace and its effect on physical and mental health and job performance is justified. The occupational stress can be considered as an accumulation of stressors. Job-related situations are considered "stressful" by most of the people at work, for example, a stressful work situation might be the one with many demands placed upon the employee, with little time for meeting them and with increased criticism from supervisors. Alternatively, occupational stress is the stress experienced by a particular individual on a particular job. Such a consideration might include whether the employee was experienced or new to the job, whether he or she routinely coped well with circumstances or was a poor survivor, and what type of personality he or she brought to the job.

Beehr and Newman (1978) outlined three categories of the symptoms of occupational stress: psychological, physical health, and behavioral symptoms. Psychological symptoms are those emotional and cognitive problems that occur under conditions of job stress. Job dissatisfaction includes disliking coming to work and finding little reason for doing well on the job. Additional psychological symptoms are depression, anxiety, boredom, frustration, isolation, and resentment. Physical symptoms are more difficult to define because, while particular work conditions have been linked to certain physical ailments and conditions, it is difficult to know how much these ailments are "caused" by the job itself versus other aspects of the worker's life. One of the most common physical health symptoms of job stress is cardiovascular disease. There significant research has been done that links stressful work conditions to the risk factors of cardiovascular diseases (Sutherland and Cooper, 1990). There is also an established link between occupational stress and gastrointestinal conditions such as ulcers. Other physical conditions that may result from ongoing occupational stress are allergies and skin diseases, sleep disturbances, headaches, and respiratory diseases Mayer (2000). Behavioral symptoms occur in two categories. The first category includes symptoms that can be said to "belong" to the worker. This group includes such behaviors as avoidance of work, increased alcohol and drug use, overeating or under eating, aggression towards fellow workers or family members, and interpersonal problems, in general. The other category includes behavioral symptoms "belonging" to the organization: absenteeism, leaving the job, accident proneness, and loss of productivity. To sum up, occupational stress can be visible to an observer with symptoms that describe an individual, such as ulcers or a depressed mood or increased hostility.

Though occupational stress initially arises from constituent factors of job and its psychophysical environment, these factors are not inherently stressors. In fact, personal characteristics of the employee and his cognitive appraisal of the job factors in the framework of his capacity and resources determine the extent of stress he experiences from a job factor of situation. However, some factors or work conditions, such as extreme heat or cold, chronic dangers, demotion, loss of job and so forth, are likely to cause stress to majority of the workers. But stress resulted from these factors also vary from one worker to another. The pressure caused from the job factors, in fact, is mediated by the personal characteristics of the focal worker. Moreover, certain psychological and behavioral specialties of the employee also become consistent sources of stress to the employee.

Although gender is not a personality characteristic, but a person's gender is part of what the individual brings to the workplace. The mediating/moderating effect of gender difference on day to day experiences of workers, particularly given the growing number of women at workplace and in jobs that have traditionally been regarded as the province of men, may be found. Hence, this changing role of women in society and in the workforce has led to greater consideration of the influence of gender on occupational stress. Smith (1979) pointed to a "subtle revolution," where the pattern of women's lives has changed from one where family responsibilities and work roles were sequential, not simultaneous, to the one where family and work responsibilities are occurring at the same time. Statistics support this contention.

Research suggests that women experience certain stressors to a greater degree with different effects than men do. Wortman et al. (1991) have considered overload of the role women professionals with preschool children. In this research, they interviewed women over a period of time concerning role conflict and role strain. One of their primary interests was in the conflict these women experienced between their work and family responsibilities. The frequency of these conflicts was striking, since the women reported such conflicts virtually daily, and the average frequency was 2–3 times a week. Their husbands, however, reported work–home conflicts once each week on an average. Interestingly, the husbands' estimates of the frequency of their wives' conflicts were significantly less than the actual frequency.

Wortman et al. also considered women's work in context of spillover. This term represents the influence of one domain on another, such as might occur when, for example, Don carries home his resentment of an argument with a boss and is bothered all evening by that. The data supported this contention, revealing that the women's job overload was associated with increased marital strain and dissatisfaction; the husbands also reported that the quality of their marriage was suffering because of the job demands of their wives.

As noted by Frankenhaeuser et al. (1991), we do not know much about women's health nor do we understand the implications of their increased involvement in the labor force on their health. In addition, research has often treated women as a homogeneous group, "averaging" across the effect of different types of occupations and jobs as well as across the different life stages of women.

Studies have suggested that a number of work setting variables like role characteristics; role ambiguity (Beehr, 1985a, Ivancevich and Matteson, 1980, Schuler, 1984), role overload (French and Caplan, 1973), role underload (Taylor, 1911), role conflict (Beehr, 1985a; Ivancevich and Matteson, 1980), job characteristics; work pace (Salvendy, 1981, Smith, 1985), repetition of work (Wallace et al., 1988), shift work (Rutenfranz et al., 1985, Monk and Tepas, 1985, Ivancevich and Matteson, 1980) and task attributes (Turner and Lawerence, 1965), interpersonal relationship (Payne, 1980, Ketz de Vries 1984, French and Caplan, 1973, Schuler, 1984, Maslach and Jackson, 1981), organizational structure (Ivancevich and Matteson 1980, Ivancevich and Donnely 1975, Schuler 1980; Cooper 1987, Ivancevich et al., 1982), organizational culture (Jick, 1985), organizational territory (Ivancevich and Matteson, 1980, French and Caplan, 1973), human resource, management practice, physical qualities and technology (Woodward, 1965, Schuler, 1977), and career development (Cartwright and Cooper, 1992, Ivancevich and Matteson, 1980) apparently influence the occupational stress.

As far as nursing profession is concerned, it can be very rewarding but is equally challenging. Ward (2014) identified some problems that nurses face universally. Some of them are of interest here:

Staffing: As healthcare costs increase, decreased number of staff nurses is often seen as the logical way to fight it. Inadequate staffing and/or increased job responsibilities both cause problems and stress for nurses. This is especially true for those staff nurses who face a variety of patient perceptions.

Interprofessional relationships: Conflicting views and feeling that nurses are being disrespected often cause problems. These conflicts in nursing relationships can arise between nurse and their patients, their coworkers, partner physicians, and/or administrators.

Patient satisfaction: Patients have needs and expectations, but unfortunately, meeting every one of them is difficult. This is especially true as healthcare conditions become more chronic, and the number of personnel decreases.

On-the-job hazards and safety: Overflowing of sharp containers and slippery floors can pose risks for staff members. Similarly, lifting heavy patients can pose a physical challenge. And since nurses do work with those patients who are "sick," there is a likelihood of contracting their illness. These hazards also include the behavior of other people. In some cases, nurses report feeling threatened by angry patients. That lack of respect and/or verbal abuse may also come from administrators, physicians, and other members of the care team.

Mandatory overtime: Due to insufficient staffing levels and/or high patient acuity, the nurses have to go on mandatory overtime.

Tendency to "ask the nurse": Nurses do not know everything about healthcare and related diagnoses. However, friends and family still feel free to call them at any given hour to ask their perspective on a symptom or prescribed medication, just because they are in the nursing profession.

Patient relationships: It is easy to develop a close relationship with the patients. When patient struggles or even dies, nurses feel their pain and they feel that loss. So maintaining a healthy, professional relationship with the patient is a must.

Advances in technology: With the growth of the Internet and smart phones, the roles and duties of nurses have changed. Documentation and databases are now mostly electronic. Using Skype to communicate is also common, as being professionals keeping abreast with these advances is essential.

Likewise, the office clerks are expected to have significantly more stressed due to their job. Office work, once considered safe, clean work, is now known to involve serious health hazards. They have to perform duties too varied and diverse to be classified in any specific office clerical occupation, requiring some degree of knowledge of office management systems and procedures. Clerical duties may be assigned in accordance with the office procedures of individual organization and may include a combination of answering telephones, bookkeeping, typing or word processing, stenography, office machine operation, and filing. Klitzman and Stellman (1989) found in their study that adverse environmental conditions, especially poor air quality, noise, ergonomic conditions, and lack of privacy, may affect satisfaction and mental health of office workers.

Pratt (1978) attempted to discover the causes of stress among teachers. Results showed that stress arose from five main areas: a general inability to cope with teaching problems; noncooperative children; aggressive children; concern for children's learning; and staff relationships. Financial deprivation in the home background was found to be positive and most significantly related to the incidence of perceived stress among teachers of all but the very youngest children; among those teaching the more deprived, stress increased with the age of children taught. A positive association was found between the amount of stress recorded and illness, as measured by the General Health Questionnaire.

During the past several years, there accumulated a mass of literature on occupational stress, which reflects a growing interest in the concerned area. Since occupational stress is not an objective phenomenon, rather it is subjectively experienced as it depends upon the individual's cognitive appraisal of the stress for agents. Reviews of researches on occupational stress suggest that it is related to a variety of individual variable (such as age, sex, self-esteem, ego level etc.), job characteristics (viz. growth, satisfaction, autonomy, salary promotions, etc.), and organizational variables (viz. work climate, feedback, suspension, coworker support, etc.). An attempt has been made for a comprehensive review of related studies conducted in the field of occupational stress.

Miles (1976) found the relationship between major role requirements and experienced role stress on the basis of data drawn from 202 research and development professionals. Measures of the role stress included various types of role conflict and ambiguity. Role conflict appeared to be more sensitive than role ambiguity to differences in Research and Development (R & D) role requirements.

Beehr (1976) investigated the situational moderators of the relationship between one organizational stress, role ambiguity, and four psychological strains: job dissatisfaction, life dissatisfaction, low self-esteem, and depressed mood. Three situational characteristics were hypothesized to moderate the relationship by reducing its strength: group cohesiveness, supervisor support, and autonomy. Group cohesiveness moderated the relationship between role ambiguity and two of the role strains, but the direction of its moderating influence was inconsistent. An explanation was offered for this result. Supervisor support showed a nonsignificant tendency to reduce the strength of the relationship between role ambiguity and role strain. Autonomy tended to moderate the relationship in the expected direction significantly and strongly.

Arsenault and Dolan (1983) investigated the relationship between job content and sources of stress and selected behavioral and attitudinal outcomes, absenteeism and performance, while controlling the differences in personality, occupation, and organizational culture. Shaw and Riskind (1983) investigated whether any consistent relationship exists between the behavioral characteristics of different jobs and the levels of various stresses experienced by the groups of employees in those jobs. Behavioral characteristics of jobs were assessed with data from the Position Analysis Questionnaire (PAQ) data bank. Correlational and regression analyses were conducted to determine the relationship between job dimension scores derived from the PAQ and 18 indices of job stress obtained from three archival sources (with job numbers of 92, 31, and 92, respectively). PAQ and stress data were matched using job titles and codes from the *Dictionary of Occupational Titles (DOT)*. Results showed a strong relationship between PAQ scores and stress data.

Fimian (1986) studied on the presence or absence of peer and administrative support in terms of the frequency and strength of stress reported by three statewide samples of special education teachers (Ns=365, 371, and 371). One ANOVA was conducted for each stress variable to determine significant differences, if any, between administrative and peer support recipients versus non-recipients. Stress variables included the strength and frequency of personal/professional stressors, professional distress, discipline and motivation, emotional manifestation, bio-behavioral manifestation, physiological, fatigue manifestation, and total stress. A majority of the group comparisons indicated stronger and more frequent stress levels for non-recipients of supervisory support than for recipients. Also, a smaller number of group comparisons indicated stronger and more frequent stress levels for non-recipients than for the recipient of peer support. Findings were generally consistent across all the three samples.

Kagan (1988) explored the relationship between teachers' cognitive styles, the kind of leadership style they prefer, and the types of occupational stress they experienced and these were explored by obtaining self-reports from 70 elementary school teachers. The teachers preferred principals who were strict rather than the relationship oriented. Preferences in leadership style appeared to be related to subjects' tendencies to perceive and evaluate situations in particular ways. Compatibility with a principal's leadership style may have depended on how closely the style matched a teacher's fundamental affective and cognitive characteristics.

Puffer and Brakefield (1989) investigated the relevance of task complexity as a moderator of the stress and coping process for a sample of 173 museum store managers. Four categories of coping responses were developed: active cognitive, active behavioral, cognitive avoidant, and behavioral avoidant. Results showed that task complexity moderates the relationship that coping had with some individual and environment antecedents as well as work outcomes. Relationship was generally stronger for simple tasks than complex ones.

Jamal and Baba (1992) studied the relationship of a shift work and department-type with employees' job stress, stressors, work attitudes, and behavioral intention. Data were collected by means of a structured questionnaire from nurses (N = 1148) working in eight hospitals in a large metropolitan city in eastern Canada. One-way ANOVA, MANOVA, and two-way ANOVA were used to analyze the data. Results generally supported the prediction that nurses working on fixed shifts were better off than nurses working on rotating shifts in terms of the dependent variables of the present study. The prediction that nurses working in non-intensive care departments were better off than nurses working in intensive care departments received mixed support at best. A few interaction effects of shift work versus department type on dependent variables were also noted. The impact of sociodemographic variables—age, marital status, cultural background (English versus French-speaking)—on the above relationships were also analyzed. Results are discussed in light of the previous empirical evidence on shift work and department-type.

Tharakan (1992) hypothesized that professional women and nonprofessional working women would differ in their job-related stress and level of job satisfaction. A sample of 90 technocrat working women (doctors, engineers, and lawyers) was compared with 90 non-technocrat working women (clerks, officers, and teachers) on these variables. The operational stress indicator scale (Cooper, 1980) was administered to measure occupational stress and job satisfaction. The relationship between occupational stress and job satisfaction was found to be significantly associated with the professional qualifications of the women. Professional working women experienced greater work-related stress than nonprofessional working women because the expectations of the former were much higher than those of the latter.

Banerjee and Gupta (1996) studied the moderating effect of social support in relation to occupational stresses and strains among 100 males and 100 females from four different occupations: police officers, advocates, doctors, and clerks. There were 25 males and 25 females in each occupation category. Caplan's Questionnaire (1979), Quinn's Questionnaire (1971) and House and Well's Questionnaire were used to collect data. Multiple regression analysis was done to test the moderating effect of social support by comparing the R^2 values of high and low social support groups split at the quartile point. Results indicate the moderating effects of social support in relation to stress and strain.

Sekhar (1996) studied job stress, job-related anxiety and helplessness, and job burnout experiences among nursing personnel from three hospitals. A group of 120 nurses, 40 each from corporate, university, and government hospitals was administered scales to measure job stress, work-related anxiety, helplessness, and burnout experiences. Results revealed that the type of hospitals differentially affected job stress and job burnout experiences. University hospital nurses scored lower on all the stress and burnout experiences than the other two groups. Further, beta coefficients calculated for number of patients nursed and dimension-wise stress and burnout experiences indicated that nurses' helplessness, depersonalization experiences, and personal accomplishment were significantly affected by the number of patients nursed.

Chand and Sethi (1997) examined the organizational factors as predictors of job-related strain. The respondents were 150 Junior Management Scale-1 Officers working in various banking institutions in the state of Himachal Pradesh. The findings show significant positive relationships between job-related strain and role overload, role conflict, and strenuous working conditions. The relationships of other organizational variables were also in the expected direction, though not significant. Role conflict, strenuous working conditions, and role overload were found to be the most significant predictors of job-related strain.

Sud and Malik (1999) studied job stress, social support, and trait anxiety among school teachers. A sample of 200 public and government school teachers was administered the Teachers Stress Survey, the Social Support Questionnaire (House and Wells, 1978), and the Social Provision Scale (Russell and Cutrona, 1986). In addition, demographic and biographic details of the teachers were collected. Correlation and multiple regression analyses showed a clear association between demographic details and social support. The close correspondence between the coworker's support and the provision of reassurance of worth ensured the moderating effect of this type of support in job-related stress situations.

Virk et al. (2001) investigated the effect of job status, age and type-A behavior on occupational stress and work motivation of nursing professionals. The sample was drawn from 295 nurses who were administered Jenkins Activity Survey (1879) to select the type-A and type-B subjects. A $2 \times 2 \times 2$ factorial design with two levels of each of the three independent measures, that is, job level and type-A behavior was used. The sample consisted of 117 nursing professionals who further responded on occupational stress index (Srivastava and Singh, 1981) and work motivation (Srivastava, 1981) test A separate analysis of variance each for the two dependent variables was employed. Results revealed that type-A behavior measure rendered significant effect on occupational stress and work motivation of the subjects. Job level and age variables also yielded significant differences in the work motivation of nursing professionals. A significant second order interactive effect of job level × type-A behavior × age was found for occupational stress. Measure of type-A behavior also interacted with job level of the subjects to affect the work motivation of the staff nurses in a significant manner.

Srivastava and Singh (2002) examined the relationship between job and life stress and health outcomes of management personnel. A sample of 200 male completed questionnaires covering occupational stress, life stress, psychosomatic health complaints (PHC), and pathogenic health habits (PHH), data on blood pressure (BP) were also collected. Job stress was significantly related to PHC and PHH. As compared to job stress, life stress was found to be a stronger predictor of health outcomes. Life stress was significantly related to higher systolic BP, PHC, and PHH.

Parikh et al. (2004) explored nurses' occupational stressors and coping mechanism. In nurses, occupational stress appears to vary according to individual and job characteristic and work family conflict. Common occupational stressors among nurses are workload, role ambiguity, interpersonal relationship, and death and dying concerns. Emotional distress, burnout, and psychological morbidity could also result from occupational stress. Nurses' common coping mechanisms include problem solving social support and avoidance. Perceived control appears to be an important mediator of occupational stress. Coping and job satisfaction appear to be reciprocally related. Shift work is highly prevalent among nurses and a significant source of stress. The effects moderating influences, coping mechanism, and risk factors associated with shift work are considered in detail here. Prophylactic and curative measures are important for nurses at both personal as well as organizational levels.

6.1.1   IMPORTANCE OF THE STUDY

Stress at work affects employee behavior in adverse ways. Some of the effects associated with stress are neuroses, coronary heart disease, alimentary conditions such as dyspepsia and ulcers, cancer, asthma, hypertension, backaches, and the use of alcohol and drugs (Beehr and Schuler, 1982). It is believed that stress can cause these problems or at least make them more severe. The list is quite broad and diverse, and that fits with Selye's (1956) description of stress as having nonspecific results on the individual. It is a part of this very generality of effects that leads to the conclusion that stress is indeed important in our everyday lives.

In terms of its effect on the economy of the nation, stress at work seems to play a very important role. Nationally, for example, the results of stress might be seen in overuse of medical and mental health facilities, or reduce Gross National Product due to increased illnesses. It has been estimated that the economic cost of peptic ulcers and cardiovascular disease, to name but two potential effects of stress are around $ 45 billion annually in the United States (Moser, 1977), and Greenwood (1978) has estimated that the cost of executive stress alone is in the billions of dollars. Most such estimates are based upon the direct costs of illnesses. In addition, it is probable that there is some additional cost due to decreased organizational effectiveness of employees who show up for work but who are operating at reduced levels of effectiveness.

6.1.2   RATIONALE OF THE STUDY

From review of existing researches, it appears that in spite of various studies on occupational stress and its correlates, some gaps still exist. Occupational stress among nurses, office clerks, and teachers was not investigated. Earlier researchers had hardly investigated these professions in relation to occupational stress. That is the reason why no study seems to explore impact of the variable on occupational stress. It was observed that previous studies have not studied occupational stress in relation to the professions of working women.

6.1.2.1   PURPOSE

To examine the impact of type of profession—nursing, teaching and office clerks, and teaching on the occupational stress of working women.

6.1.2.2   HYPOTHESES

There will be significant difference in the occupational stress scores of women respondents working in different type of professions.

6.1.2.3   VARIABLE STUDIED

Dependent variable—Scores on occupational stress index
Independent variable—Type of profession

6.2   METHODS

6.2.1   PARTICIPANTS

In this study, 270 working women belonging to the professions of teaching, nursing, and office clerk participated. They were drawn from eastern districts of Uttar Pradesh and their age range was 30–45 years (Mean=39.73 years, and SD=4.67).

6.2.2   TOOLS

Occupational stress index (OSI) by Srivastava and Singh (1981) has 46 items. This scale comprises 12 subscales: role overload (six items), role ambiguity (four items), role conflict (five items), unreasonable group and political pressure (four items), responsibility for persons (three items), under participation (four items), powerlessness (three items), poor peer relations (four items), intrinsic empowerment (four items), low status (three items), strenuous working conditions, (four items) and unprofitability (two items). The tool may conveniently be administered to the employees of every level of the operation in context of industries or other non-production organizations. The split-half reliability index and Cronbach's alpha coefficient for the scale was 0.935 and 0.90, respectively. The validity of OSI was determined by computing coefficients of correlation between the scores on OSI and various measures of job attitudes and job behavior. The coefficient of correlation between scores on OSI and the measures of job involvement (Lodhal and Kejner, 1965), work motivation (Srivastava, 1984), ego strength (Hasan, 1970), and job satisfaction (Pestonjee, 1973) were found to be -0.56 ($N=225$), -0.44 ($N=200$), -0.40 ($N=205$) and -0.51 ($N=500$), respectively. The correlation between the scores on the OSI and measures of job anxiety (Srivastava, 1974) was found to be 0.59 ($N=400$).

6.2.3   PROCEDURE

Prior to actual administration of the test for collecting data for the study, a preliminary trial of the tool was conducted. The purpose of this was to determine the time needed for administration of the test and to review the testing procedure for any unforeseen difficulties. As the sample (270) of present study consisted of three sorts of women respondents belonging to nursing, teaching, and office clerks, in this way, the respondents were, first of all, categorized into three categories on the basis of type of profession. Then they were administered OSI.

6.2.4   STATISTICAL ANALYSIS

The obtained raw scores on OSI as function of the type of professions were statistically analyzed herewith the purpose clarifying and explaining the problem raised and hypotheses formulated.

The statistical measures employed in the study were:

1. Mean values were calculated for graphical presentation of the data OSI related to different type of professions.
2. Mean, median, mode, standard deviation, quartiles (Q_1 and Q_3), percentiles (P_{10} and P_{90}), skewness, and kurtosis were computed as the measure of variability.
3. One-way analysis of variance has been employed to explore the impact of profession type on occupational stress.

6.2.5 DISTRIBUTION OF SCORES

Prior to the application of statistical measures on the raw data, its interpretation and assessment of the nature of scores seems to be a plausible step. There are definite laws of existence of variability in a population that form the basis of hypothesis of normal probability. Any deviation from it can be best interpreted in the form of variability measures: skewness and kurtosis. In such samples, the frequency polygon would not exhibit unimodal symmetry. Odell (1957) prescribed the criterion specifically; he said that the positive value of skewness indicates that the mean of the data exceeds both, their median and mode, and that there is greater bunching of the measures in the direction of the low scores or greater extension or "tailing out" in that of the high. A maximum value is ±3, but a greater than one indicates extreme skewness or asymmetry (p. 65).

With regard to the interpretation of the value of kurtosis, Odell says: "If the measure of kurtosis is 0.263, the distribution is platykurtic, if smaller it is leptokurtic. It is always positive and ranges from 0.00 to 0.50. If it is 0.00 at least the middle 50% of the cases, but not the middle 80% are bunched at a single point, if it is 0.3125, the distribution or at least the middle 80% of it, may form a perfect rectangle, if it is 0.50 all the cases between the tenth and first quartile are bunched at one point" (p. 66).

According to Garrett (1965), the standard error of skewness is not considered very dependable for setting up the acceptable limits within chance of fluctuation (p. 101).

Frequency distribution based on raw scores regarding occupational stress is given in the Table 6.1.

TABLE 6.1   Frequency Distribution of Occupational Stress Score.

Class interval	(f)
130–134	19
125–129	56
120–124	125
115–119	55
110–114	15
	$N = 270$

Mean $= 122.17$
Median $= 122.1$
Mode $= 122.02$
S.D. $= 4.78$
$P_{10} = 115.59$
$P_{70} = 128.78$
$Q_1 = 119.27$
$Q_3 = 125.17$
$Q_D = 5.89$

Keeping in view the above consideration and significant values of skewness and kurtosis at 0.01 level, the description of dependent variable is given below:

TABLE 6.2   Mean, Median, Mode, Standard Deviation, Skewness, and Kurtosis for The Measure of Occupational Stress.

Variable	Mean	Median	Mode	SD	Sk	Ku	Significance
Occupational Stress	122.17	122.1	122.02	4.78	0.042	0.446	NS

Skewness $= 0.235$ significant at 0.01 level
Kurtosis $= 0.467$ significant at 0.01 level

In the light of results given in the Table 6.2, it is obvious that the values for mean, median, and mode did not show wide departure; an indication that the distribution is approximately normal in the case of the total sample.

The values of kurtosis and skewness are within the limits of chance fluctuations. The nature of distribution as viewed from the value of kurtosis for the total sample with the values of skewness is found to be normal (see Fig. 6.1). This suggests that the proximity of the distributions conform to the normal sample.

Following facts emphasized that the sample under study is normally distributed.

1. Table 6.2 exhibits that there is a very slight difference in the mean, median, and mode values of the frequency distribution of Occupational Stress scores (Table 6.1), thus indicating that Occupational Stress is almost normally distributed in the selected sample.

2. The frequency distribution curve (Fig. 6.1) is almost a normal curve, as the skewness and kurtosis is found in the curve is negligible.

3. When the sum of positive and negative deviations is not zero, the distribution may be skewed, however, it is not so in this frequency distribution (Table 6.1).

4. When the difference between mean and Q_1 (122.17−119.27=2.9) and mean and Q_3 (122.17−125.17=3.00) are not at equal distance; the distribution is skewed. However, in this study Q_1 and Q_3 are almost at equal distance.

5. A positive skewness is found when the value of Q_3 median exceeds the value of median Q_1. If the value of Q_3 median is less than the value of median Q_1, negative skewness is found. If the difference between these two values is zero, there is no skewness in the sample. The difference is 0.24 between these values hence, the distribution is approximately normal.

6. Table 6.2 shows a kurtosis of 0.446, however, the value of skewness is not significant at 0.01 level of confidence.

7. The value of skewness =0.042 is also insignificant.

In the light of above statistical analysis, it can be ascertained that the sample selected for the study of occupational stress in relation to their type of profession is a normal sample. Whatever slight deviation was found is just by chance and hence, any study conducted on such sample can be attributed to a larger population (See Fig. 6.1).

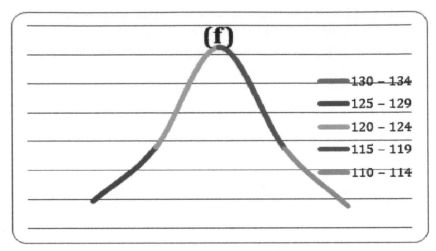

FIGURE 6.1   Normal distribution curve for occupational stress scores of 170 respondents of three professional groups.

6.3   RESULTS

For the comparison of occupational stress score among working women from three different professional groups, Mean and SD values of occupational stress scores as the function of type of profession was calculated and presented in the Table 6.3.

TABLE 6.3   Mean and SD Values Obtained on Occupational Stress Index (OSI) by Women Respondents as Function of their Type of Profession.

Type of profession	Mean	N	SD
Teaching	119.96	90	5.7
Nursing	123.57	90	5.6
Office clerks	122.03	90	4.5

Table 6.3 reveals that occupational stress of women respondents belonging to three different professions. The nursing group was found to be the more stressed group (M = 123.57) because higher score on OSI indicates higher stress in individual. The office clerks and teaching groups had a mean value of 122.03 and 119.96 respectively. Teaching profession emerged as the least stressed group with the lowest mean value meaning

thereby that Teaching profession has exhibited least stress towards their occupation (Fig. 6.2).

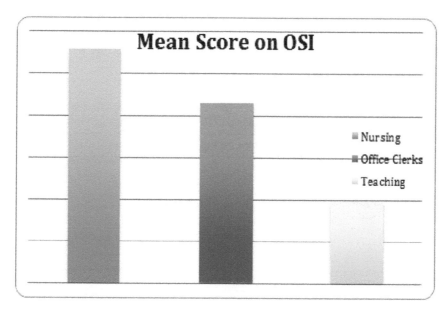

FIGURE 6.2   Bar graph showing the grand mean performance of women respondents belonging to three different professional groups.

To test the significance of difference in these mean values as exhibited in Table 6.3, one-way ANOVA has been employed and presented in Table 6.4.

TABLE 6.4   Summary of Analysis of Variance.

Occupational stress	Sum of squares	df	Mean square	F	Sig.
Between groups	591.252	2	295.626	10.451	0.000
Within groups	7552.822	267	28.288		
Total	8144.074	269			

With a view to see whether these characteristic of the data evinced by usual inspection is statistically significant.

According to Garrett (1965), *F*-value furnishes a comprehensive or overall test of the significance of the differences among means. But a significant F value does not tell which means differ significantly but that of least one is reliably different from some others. If F is not significant, there is no reason for further testing, as none of the mean differences will be significant. But if F is significant we may proceed to test the separate differences by the post hoc test. Thus, to check the groups among which the significant difference in occupational stress existed, the post hoc test has been made (see Table 6.5).

TABLE 6.5 Multiple Comparisons of Difference in Mean Scores Obtained on OSI of Women Respondents Belonging to Teaching, Office Clerks, and Nursing Professions.

(I) Type of profession	(J) Comparison group	Mean difference (I–J)	Sig.
Teaching	Office clerk	−2.078*	0.025
	Nursing	−3.611*	0.000
Office clerk	Teaching	2.078*	0.025
	Nursing	−1.533	0.131
Nursing	Teaching	3.611*	0.000
	Office clerk	1.533	0.131

*The mean difference is significant at the 0.05 level.

Table 6.5 presents the comparative observation of mean scores on OSI of working women belonging to teaching, nursing, and office clerk professions. The mean difference between the groups of nursing and office clerks was found insignificant on 0.05 level. The difference showed were may be due to some fluctuation of chance factors. The mean differences between teaching–office clerks and teaching–nursing were found significant.

6.4 DISCUSSION

The obtained result from the present study permits us to answer the question regarding the impact of type of professions on the occupational stress of working women. It was found that the working women serving in nursing profession scored comparatively higher marks on OSI then followed by the office clerks and lastly the teachers meaning that the women respondents belonging to the nursing professions were more stressed towards their jobs than office clerks and teachers. To see the significance of mean difference, one-way ANOVA was applied. Obtained F value was found highly significant at respectable level of confidence. It does mean that there are significant differences among the scores of working women respondents belonging to nursing, teaching, and office clerks. Obtained result is confirming the hypothesis that "There will be significant difference in the occupational stress scores of women working in different type of professions." The result indicates that the profession type is an important factor that affects the degree of the occupational stress of working women.

Possible reason for such trends in the observed pattern of results may be speculated in terms of gender role differences (Smith, 1979), social role of women across different religio-cultural settings and social support (Banerjee and Gupta, 1996)), nature of work responsibilities (Arsenault and Dolan, 1983), shift work (Jamal and Baba, 1992) and so forth. Due to the problem and difficulty in staffing, interprofessional relationships, patient satisfaction, on-the-job hazards and job safety, mandatory overtime, "ask the nurse", patient relationships, advances in technology, certification and so forth, the nursing women have more chances to be stressed toward their jobs (Ward, 2014).

Nursing women have to work with the patients in different socio-cultural settings; many times they have been in between the community health centers and field as required from time to time. They have to work in shifts, too. That is why the nursing women are expected to have more occupational stress. As the office clerks were found to be more stressed towards their job than teachers. The office clerks have to perform their duties too varied and diverse to be classified in the office clerical occupation, requiring some degree of knowledge of office management systems and procedures. The adverse environmental conditions, especially poor air quality, noise, ergonomic conditions, and lack of privacy, may affect satisfaction and mental health of office workers (Klitzman and Stellman, 1989). The reason behind the stress among teachers is probable as general inability to cope with teaching problems; noncooperative children; aggressive children; concern for children's learning; and staff relationships (Pratt, 1978).

Intervention programs, especially more specific to women may be developed by considering the nature of the results as obtained in the present study.

6.5   CONCLUSION

The results of present investigation indicate that:

1. The level of occupational stress of working women is influenced by their type of profession.
2. Working women belonging to nursing profession were found to be more stressed towards their job then followed by the office clerks and teachers.

KEYWORDS

- occupational stress
- type of profession
- nurses
- office clerks
- teachers

REFERENCES

Arsenault, A.; Dolan, S. The Role of Personality, Occupation and Organization in Understanding The Relationship Between Job Stress, Performance and Absenteeism. *J. Occup. Psychol.* **1983,** *56,* 227–240.

Banerjee, U.; Gupta, H. N. Moderating Effect of Social Support in Occupational Stress Strain Relationship. *J. Indian Acad. Appl. Psychol* **1996,** *22*(1–2), 27–34.

Beehr, T. A. Perceived Situational Moderators of The Relationship Between Subjective Role Ambiguity and Role Strain. *J. Appl. Psychol.* **1976,** *61*(1), 35–40.

Beehr, T. A. "Organizational Stress and Employee Effectiveness: A Job Characteristics Approach." In *Human Stress and Cognition in Organizations;* Beehr, T. A., Bhagat, R. S., Eds.; John Wiley and Sons: New York, 1985a.

Beehr, T. A.; Newman, J. E. Job Stress Employee Health and Organizational Effectiveness: A Fact Analysis Model and Literature Review. *Pers. Psychol.* **1978,** *31,* 665–669.

Beehr, T. A.; Schuler, R. S. Current and Future Perspectives on Stress in Organizations. In *Personnel Management: New perspectives;* Rowland, K. M., Ferris, G. R., Eds.; Allyn and Bacon: Boston, 1982.

Caplan, R. D. The Family as a Support System. In *Support Systems and Mutual Help: Multidisciplinary Explorations;* Caplan, G., Killies Eds.; Grune and Strattom: New York, 1979.

Cartwright, S.; Cooper, C. L. *Mergers and Acquisitions: The Human Factor;* Butterworth Heinemenn: Oxford, 1992.

Chand, P.; Sethi, S. Organizational Factors in the Development of Work Stress. *IJIR* **1997,** *32,* 4.

Cooper, C. L. *Stress Research: Issues for the Eighties;* John Wiley: New York, 1980.

Cooper, C. L. The Experience and Management of Stress: Job and Organizational Determinants. In *Occupational Stress and Organizational Effectiveness;* Reley, A. W., Zaccaro, S. J., Eds.; Praeger: New York, 1987.

Fimian, M. Social Support and Occupational Stress in Special Education. *Exceptional Child* **1986,** *52*(5), 436–442.

Frankenhaeuser, M.; Lundberg, U.; Chesney, M.; Eds. *Women, Work and Health;* Plenum: New York, 1991.

French, J. R.; Caplan, R. D. Organizational Stress and Individual Strain. In *The Failure of Success;* Marrow, A. J., Ed.; AMACOM: New York, 1973.

Garrett, H. E. *Statistics in Psychology and Education;* Vakils, Feffer and Simons Ltd.: Bombay, 1965.

Greenwood, J. W. Management Stressors. In *Reducing Occupational Stress;* Cincinnate: NIOSH Research Report, 1978.

Ivancevich, J. M.; Donnelly, J. H. Relations of Organizational Structure to Job Satisfaction, Anxiety—Stress and Performance: Measure Research and Contengencies. *Administrative Sci. Q.* **1975,** *20,* 272–280.

Ivancevich, J. M.; Matteson, M. T. *Stress and Work. A Managerial Perspective;* Scott Foresman: Glenview, IL, 1980.

Ivancevich, J. M.; Mattenson, M. T.; Preston, C. Type a Behaviour and Physical Well Being. *Acad. Manage. J.* **1982,** *25,* 373–391.

Jamal, M.; Baba, V. V. Shiftwork and Department-Type Related to Job Stress, Work Attitudes and Behavioural Intentions: A Study of Nurses. *J. Organ. Behav.* **1992,** *13,* 449–464.

Jick, T. D. As The Ax Falls: Budget Cuts and The Experience of Stress in Organizations. In *Human Stress and Cognition in Organizations;* Beehr, T. A., Bhagat, R. S., Eds.; John Wiley: New York, 1985.

Kagan, D. M. Inquiry Mode, Occupational Stress and Preferred Leadership Style Among American Elimentary School Teachers. *J. Soc. Psychol.* **1988,** *129*(3), 297–305.

Ketz de Vries, M. F. R. Organizational Stress Management Audit. In *Handbook of Organizational Stress Coping Strategies Cambridge;* Sethi, A. S., Schuler, R. S., Eds.; M.A. Ballinger: Cambridge, 1984.

Klitzman, S.; Stellman, J. M. The Impact of The Physical Environment on the Psychological Well-Being of Office Workers. *Soc. Sci. Med.* **1989,** *29*(6), 733–742.

Lodhal, T. M.; Kejner, M. The Definition and Measurement of Job Involvement. *J. Appl. Psychol.* **1965,** *49,* 24–33.

Maslach, C.; Jackson, S. E. The Measurement of Experienced Burnout. *J. Occup. Behav.* **1981,** *2,* 99–113.

Mayer, E. A. The Neurobiology of Stress and Gastrointestinal Disease. Gut **2000,** *47,* 861–869.

Miles, R. H. Role Requirements as Sources of Organizational Stress. *J. Appl. Psychol.* **1976,** *61*(2), 172–179.

Monk, T. H.; Tepas, D. I. Shift Work. In *Job Stress and Blue Collar Work;* Cooper, C. L., Smith, M. J., Eds.; John Wiley: New York, 1985.

Moser, M. A Major Controllable Public Health Problem—Industry can Help. Occupational Health Nursing, August, 1977, 19–26.

Odell, W. *A First Course in Educational Statistics;* W.M.C. Brown Company: Bubuque, Iowk, 1957.

Parikh, P.; Taukari, A.; Bhattacharya, T. Occupational Stress and Coping Among Nurses. *J. Health Manage.* **2004,** *6,* 2.

Payne, R. Organizational Stress and Support. In *Current Concerns in Occupational Stress;* Cooper, C. L., Payne, R., Eds.; John Wiley: New York, 1980.

Pratt, J. Perceived Stress Among Teachers: The Effects of Age and Background of Children Taught. *Educ. Rev.* **1978,** *30,* (1).

Puffer, S. M.; Brakefield, J. T. The Role of Task Complexity as a Moderator of the Stress and Coping Process. *Hum. Relat.* **1989,** *42*(3), 199–217.

Putt, A. M. One Experiment in Nursing Adults with Peptic Ulcers. *Nurs. Res.* **1970,** *19,* 484–494.

Quinn, R. P.; Seashore, S.; Mangione, I. *Survey of Working Conditions,* U.S. Government Printing Press, 1971, North Capitol St. NW Washington, D.C.

Rutenfranz, J.; Haider, M.; Kollar, M. Occupational Health Measures for Night Workers and Shift Workers. In *Hours of Work: Temporal Factors in Work Scheduling;* Folkard, S., Monk, T. H., Eds.; John Wiley: Chichester, 1985.

Salvendy, G. Classification and Characteristics of Paced Work. In *Machine Pacing and Occupational Stress;* Salvendy, G., Smith, M. J., Eds.; Taylor and Francis: London, 1981.

Schuler, R. S. Role Conflict and Ambiguity as a Function of the Task Structure—Technology Interaction. *Organ. Behav. Hum. Perform.* **1977,** *20,* 60–74.

Schuler, R. S. Definition and Conceptualization of Stress in Organizations. *Organ. Behav. Hum. Perform.* **1980,** *24,* 115–130.

Schuler, R. S. Organizational Stress and Coping: A Model and Overview. In *Handbook of Organizational Stress Coping Strategies;* Sethi, A. S., Schuler, R. S., Eds.; Ballinger: Cambridge, M.A., 1984.

Sekhar, S. F. C. Job Stress and Burnout Experiences Among Nurses from Different Hospitals. *Manage. Labour Stud.* **1996,** *21*(2), 114–120.

Selye, H. *The stress of life;* McGraw-Hill: New York, 1956.

Shaw, J. B.; Riskind, J. H. Predicting Job Stress Using Data from The Position Analysis Questionnaire. *J. Appl. Psychol.* **1983,** *68,* 253–261.

Smith, M. J. Machine-Paced Work and Stress. In *Job Stress and Blue Collar work;* Cooper, C. L., Smith, M. J., Eds.; John Wiley: New York, 1985.

Smith, R. E. The Subtee Revolution. The Urban Institute: Washington, D.C., 1979.

Srivastava, A. K. An inquiry into relationship between job anxiety and job satisfaction. Unpublished doctoral dissertation, Department of Psychology, B.H.U., Varanasi, 1974.

Srivastava, A. K. *Construction and Standardization of Employees Motivation Schedule. Department of Psychology;* Banaras Hindu University: Varanasi, 1984.

Srivastava, A. K.; Singh, A. P. Construction and Standardization of an Occupational Stress Index. A pilot Study. *Indian J. Clin. Psychol.* **1981,** *6,* 133–135.

Srivastava, U. R.; Singh, A. P. Relationship of Job and Life Stress to Health Outcomes Among Indian Managerial Personnel. *Soc. Sci. Int.* **2002,** *18*(1), 47–57.

Sud, A.; Malik, A. K. Job Related Stress, Social Support and Trait Anxiety Among School Teachers. *J. Indian Acad. Appl. Psychol.* **1999,** *25*(1–2), 51–55.

Sutherland, V. J.; Cooper, C. L. *Understanding Stress: A Psychological Perspective for Health Professionals;* Chapman and Hall: London, 1990.

Taylor, F. W. *Principles of Scientific Management;* Horper and Row: New York, 1911.

Tharakan, P. N. O. Occupational Stress and Job Satisfaction Among Working Women. *J. Indian Acad. Appl. Psychol. 18*(1–2), 37–40.

Turner, A. N.; Lawerence, R. P. Industrial Jobs and Worker. Harvard University Graduate School of Business Administration: Cambridge, M.A., 1965.

Virk, J.; Chhabra, J.; Kumar, R. Occupational stress and work motivation in relation to age, job level and type—A behaviour in nursing professionals. *Journal of the Indian Academy of Applied Psychology,* **2001,** *27*(1–2), 51–55.

Wallace, M. Levens, M.; Singer, G. Blue Collar Stress. In *Causes, Coping and Consequences of Stress at Work;* Cooper, C. L., Payne, R., Eds.; John Wiley: New York, 1988.

Ward, J. Nine Common Problems in the Nursing Profession. nursetogethor.com. 2014. http://www.nursetogether.com/9-common-problems-nursing-profession#comments.

Woodword, J. *Industrial Organizations: Theory and Practice;* Oxford University Press: London, 1965.

Wortman, C.; Biernat, M.; Lang, E. Coping with Role Overload. In *Women, Work and Health;* Frankenhaeuser, M.; Lundberg, U., Chesney, M., Eds.; Plenum Press: New York, 1991.

CHAPTER 7

KNOWLEDGE MANAGEMENT TO LEARNING TO INNOVATION IN ORGANIZATIONS: THE CRITICAL ROLE OF HUMAN RESOURCES

NAVDEEP KAUR KULAR

Vedatya Institute, Gurgaon, India, E-mail: novikular@rediffmail.com

CONTENT

ABSTRACT

The size and scale of change in business environment have been unprecedented in the recent past. Globalization and liberalization have increased competition. Technology has permeated every sphere of business dimension linking internal and external stakeholders. The disruptive innovations in many industries have challenged the established ways of work and fostered new business models. The nature of workforce and the work itself is changing. Organizations can no longer remain static in this dynamic environment. The organizations should be able to access requisite resources easily and promptly to harness advantages offered by the marketplace. The knowledge and skills of the workforce need to be in tandem with the requirements of the business. The only course that would lead to sustainable growth is to make the organization capable of learning. Employee learning has become a continuous process in the work environment. The organization itself needs to be able to change its form and structure in a short duration of time. This paper builds upon the knowledge-based view of the organization leading to learning and innovation. It explores the critical role of HR in organizations for the future to build capability for generation and implementation of innovative business models. Literature was reviewed and round table and focus group discussions were conducted to arrive at practical implications.

7.1   INTRODUCTION

Organizations can no longer remain static in the dynamic environment which is bringing in profound change in the way business is largely conducted. Globalization has led to the interconnected world and has brought in new challenges. The value of merchandise trade and trade in services has doubled during the decade spanning 2005–2015 (World Trade Organisation, 2016). The liberalization has increased competition in every sphere of the economy. The slowdown in China, exchange rate volatility, recession in Brazil, decrease in prices of oil, and other commodities are some of the recent challenges being faced by the world. The value of the merchandise exported fell to $ 16 trillion in 2015, from $ 19 trillion in the previous year (World Trade Organisation, 2016) Technology has permeated all business dimensions linking internal and external stakeholders. Technological advancements make it imperative for the organizations to embrace and upgrade technology at regular intervals. Customer tastes and preferences are changing. The volatile and uncertain environment makes its forecasting and analysis challenging. The organizations cannot afford to spend undue time on analysis of multiple interlinked variables in the environment before putting a plan into action. The experience and gut feel of the managers should guide the decisions taken by the organizations.

In the knowledge economy, competition can emerge from any corner of the world and challenge the market leaders and established ways of doing work. The short product life cycles and high rate of obsolescence threaten the existence of many existing businesses. Since 2000, the digital disruption has been the cause of bankruptcy, acquisition, and closure of 52% of the Fortune 500 companies. Internet is connecting more and more things on the planet; the number of interconnected things, including sensors and radio frequency identification chips will reach up to 50 billion by 2020 (Ernst and Young, 2015). Till date the potential connectivity is only 1%. Many industries have gone into decline and others have converged to remain relevant. The music industry has come a long way from long playing records, cassette tape recorders, Walkmans, mp3 players, iPods and finally, to online distribution of music. The convergence of computing, mass media, camera, telephone, and watch industries have posed challenges for many firms. The internet-based calling systems have siphoned off the profits of telecom industry.

The smart phones have become ubiquitous. The cost of technology is showing a continuous downward trend with increased functionality and memory. The smart display features make the devices attractive to both the employees and the customers. The ecosystem of app development is gearing itself towards mobiles. The connectedness poses new challenges and offers new opportunities. The employees can remain connected to the organization 24/7 through various devices. The employees have access to the organizational resources outside of office hours. The organizations need to alter ingrained ways of work to make way for teleworking and virtual working teams. New policies for work practices need to be crafted to harness the potential of the employees. Who all are permitted to work from distance? What technologies need to be put into place? Who bears the cost of the devices to be used? How to ensure the security of organizational data? How to bring in social aspects in virtual work? How to support virtual workers? How to measure the virtual work done?

It is the human resources (HR) that deploy all the other resources in the organization to achieve competitive advantage in the marketplace. The HR whether at the leadership level, managerial level, or any other level plays an important role in success of an organization. As the environment is changing at the fast pace and organizations are charting through unfamiliar territories, it is imperative for organizations to be proactive and build capabilities to avail opportunities as and when they arise. The knowledge and skills of the workforce need to be in tandem with the requirements of the business.

The workforce in organizations consists of individuals from different generations, diverse backgrounds, and ethnicities. Each group brings its own attitude, behavioral processes, and work methods. Gen Y is entering the workforce in large numbers. There is a distinct difference in the way this generation thinks and acts, generates and consumes, stores and processes, and most importantly learns and earns. Lifelong employment has become a thing of the past. The attrition rates in organizations are heading northwards as employee loyalty is diminishing. The organizations require new techniques to engage and retain the digital generation. The 21st century is the century of the women. The women are gathering large number of degrees and are joining the workforce. The next two decades will see the entry of 1 billion women in the workforce (The Economic Intelligence Unit Limited, 2015). Research shows that boardrooms that are more balanced in terms of gender show higher profitability. The women empathize more and are able to build rapport with other team members. They bring a new perspective to the organization. Cross country migrations are increasing and workers are being hired across borders.

The availability of resources has become more democratic in nature. The organizations can no longer run set training and development modules but need to personalize content to the requirements of the individual employees. The onus of development has been shifted from the HRD to the line managers and employees themselves. Employee learning has become a continuous process in the work environment.

7.2   KNOWLEDGE MANAGEMENT

Knowledge management has its roots in the use of information technology for storage, analysis, and distribution of information through databases, intranets, and intelligence systems. The digital assets of organizations are increasing and the organizational design has become digitally embedded. Recent practices in knowledge management have focused on the organizational learning aspects. Knowledge management comprises all the activities related to knowledge creation, transfer, and utilization. Knowledge management is concerned with the organizational processes and practices aimed at generation of value through knowledge. The creation, integration, and transformation of the knowledge play an important role in providing competitive advantage to the firm. Knowledge-based view of the firm recognizes knowledge as a key resource and appreciates the value creation through deployment of knowledge. The value can be appropriated by offering goods and services in the marketplace through utilization of this knowledge. Inkinen et al. (2015) have divided the knowledge management practices into ten types and provided validated scales for measurement of each.

Knowledge is being created at an exponential pace. It is important to gather and integrate new knowledge in the firm from all external and internal sources. The role of human interactions and mutually supportive relationships are important in building a "knowledge ecosystem" (Miller, 2015). The transfer of knowledge takes place through communication and observation. The cognitive dimension of knowledge transfer forms an important parameter of the knowledge sharing platform. The knowledge of employees needs to be integrated into organizational processes. The capability of knowledge in the organization needs to be enhanced. Also the knowledge generated in one part of the organization needs to be disseminated in other parts of the organization. The transfer of knowledge needs to be smoothened. The best practices need to be recognized and assimilated in the organization. The knowledge diffusion in the firm takes place through its embodiment in the products and services. The replication of knowledge within the firm needs to be eased. The explicit knowledge can be replicated easily and at low cost through digital devices. The transfer of tacit knowledge is slow and expensive. It requires intuition and practice to transfer tacit knowledge. On the other hand, the external replication of knowledge needs to be limited to safeguard the competitive advantage. Knowledge boundaries need to be built to keep the knowledge of the firm hidden deep in the embedded processes of the firm. The complexity and opaqueness of knowledge is important to shield it from imitation by the competitors. The knowledge boundary construction and its protection are important parameters of knowledge management.

The activities related to enhancement of the organizational knowledge are termed knowledge exploration whereas those related to deployment of knowledge are known as knowledge exploitation. Knight and Harvay (2015) studied the exploration versus exploitation paradox and found that in creative organizations, the organization processes support the balance between the two forms to reap synergistic benefits. The structural and contextual organizational forms support and sustain the generations of new ideas and standardization of processes simultaneously.

Knowledge management is important not only in commercial firms but also in not-for-profit organizations. The not-for-profit organizations contribute substantially to the economy and have to compete with each other for funding. The organizations need to utilize their limited resources judiciously. Downes and Marchant (2016) found out that face-to-face transfer of knowledge was most prevalent method in the Australian community service organizations. Staldier and Fullagar (2016) appreciated the cultural context that facilitated the formal and informal knowledge transfer practices enabling collaboration and trust. Knowledge management programs under a strong leader with local champions aligned with the daily work routines help in organizational culture change (Corfield and Paton, 2016).

Sutton (2015) found the lack of awareness and use of contemporary business improvement tool namely "Human Sigma" in tourism organizations. Organizations perceive several barriers to fully engage with tools and techniques for business improvement. The researcher found that critical success factors for implementation of business improvement tool are often related to customers and not employees.

7.3   LEARNING

Learning can happen at various levels, for example, individual, group, organizational and inter-organizational. In times of uncertainty, organizations should experiment and react to external environment rather than having set plans for long durations of time. The learning organization can create, acquire, and transfer knowledge and can change its behavior according to new insights. In learning organizations people are expanding their capacity to individually learn and create and as a group are also learning how to learn (Senge, 1990). The expansive learning concept proposed by Engestorm (2015) focuses on learning of new forms of activity and collective transformation through the process of questioning, analysis, modeling, testing, implementing, reflecting, and generalizing. The organizations go through a stage of high growth and maturity after inception. Tam and

Gray (2016) have demonstrated that there are varied needs for learning in different stages of the organizational lifecycle in small and medium enterprises. Brink and Madsen (2015) found that managers from small and medium enterprises employ action oriented, collaborative and cross disciplinary approach to learning which helps in overcoming business challenges. The social approach which integrates learning from large-scale players and other peers enhances the practical use and value of learning.

A peer-to-peer learning approach based on trans-organizational development has been proposed by O'Neill and Bent (2015). The senior executives from number of businesses and organizations share and learn from each other and dispel their predispositions to respond in an ingrained manner. This is an effective method for individual as well as organizational change.

Learning is inherent in the manager's daily tasks. The line managers who adopt a coaching approach play a central role in workplace learning (Campbell and Evans, 2016). Different types of learning-conducive climate can be created through the manager's interventions. Doos et al. (2015) identified that the interventions of the middle managers in the daily tasks of subordinates can be of two different types: social and organizational structures. These are impacted based upon the type of intervention used. The subordinates can be given fixed objectives with a set pattern of interpretation, and stringent controls can be imposed that reduces discretion of all. Alternatively, employee discretion can be increased and a learning-conducive climate can be created. The involvement of all leads to emerging capabilities of organizational members.

Lancaster and Milia (2015) studied the characteristics in an organization that support learning as perceived by employees. The organization's leadership, building relationships, openness to new ideas, learning with peers, sharing, coaching, and reflection are important characteristics of a learning environment. A learning culture is created when employees from different functional areas and regions learn together. The encouragement provided by the organization makes employees self-aware and self-confident to apply the learning in real life situations at work. Smith and Martin (2014) found a strong association between being professional and lifelong learning and reflection. Helyer (2015) highlights the importance of reflection in personal and professional learning as it provides structure in order to make sense of learning and embed it in practice for constant innovation. For individuals to develop on an ongoing basis, reflection plays a crucial role. Reflection enables individuals to modify responses to current scenario on the basis of observations, articulations, and theorization.

Haase et al. (2015) showed that the concepts of organizational learning and intrapreneurship are interrelated. The nurturing and exploitation of creative opportunities within large organizations are termed as intrapreneurship. The development of intrapreneurial teams, cross-fertilization of ideas through different alliances, healthy rivalry between intrapreneurial groups, and grant of ownership rights to innovation are some of the policies that support intrapreneurship. Organizational culture change has been identified as an important supportive variable for establishment of a learning organization by Limwichitr et al. (2015). Creative organizational climate has a strong association with organizational resilience and innovation (Mafabi et al., 2015). The communities of practice can act as learning and development spaces and help diffuse organizational culture and cope with transitions (Impedovo and Manuti, 2016). The redefinition of mental models and sense making guides the individual and organizational actions.

Ekanem (2015) highlighted the gender differences in entrepreneurial learning. The male entrepreneurs utilized double-loop learning processes and tended to depart from the industry norms while the female entrepreneurs utilized single-loop process and focused on the routinized learning. Workplace learning is not the prerogative of the young only. The investigation by Warhurst and Black (2015) found that later-career managers are learning extensively at the workplace through challenges encountered and incidental experiences. Most of the occupational learning for later-career managers takes place on the job through interactions with teams, managers, and external stakeholders.

Organizations should display strategic foresight and learn not only from the successes but also from the failures of other organizations. Amankwah-Amoah and Zhang (2015) demonstrate through illustrative cases that resistance to change, senior management hubris, and overreliance on external consultants can lead to organization closure.

7.4   INNOVATION

The deployment of knowledge management practices is a significant driver of innovation in organizations (Inkinen, 2016). The innovation capability of the firm is enhanced through knowledge sharing processes. Knowledge needs to be diffused in the form of innovation in products and services in order to reap benefits in the market place. Knowledge needs to be protected in the marketplace to retain and sustain the competitive advantage of the organization. The technological advancements of the recent years have led to fast obsolescence of knowledge and emergence of new knowledge. The boundaries around knowledge are continuously blurring as multiple industries are merging with one another and new ones are emerging.

The knowledge of the firm needs to be updated and managed on an ongoing basis to provide competitive advantage. The multiple databases need to be integrated and synchronized for quick access and use. The storage, use, access, and updation needs to be regulated. The knowledge management systems of the organization should provide impetus for quick collaboration and creation of new knowledge. At the same time, the knowledge of the firm needs to be safeguarded against any pilfering and misuse. The new knowledge needs to be diffused and embedded deep in the processes of the organization. Paraponaris and Sigal (2015) have carried out the meta-analysis of the knowledge literature and found that the focus has now shifted from knowledge boundaries to collaborative processes of creation. The knowledge management systems of the organizations alter the form that boundaries take to facilitate expansion in knowledge. The knowledge sharing process depends upon the context in which interaction happens in an organization and the cognitive dimension of exchange. The boundaries have to be crossed to create new knowledge, assimilate it and diffuse it in the form of innovation.

Open innovation is the way forward. In open innovation the borders of the organization are opened up and inputs are taken from all stakeholders. Data sharing is an important component to feed forward the innovation process. The policy to share data with complex networks within and outside the organization aids in this process. The conventional managerial and steering tools become redundant. The meaning and identity is created in this collaborative approach through continuous intervention to develop and change context (Ollila and Ystrom, 2015). The organizations and communities through interactions with different actors inside and outside the borders in various contexts expand their horizons (Impedovo and Manuti, 2016). The leading companies collaborate with customers and business partners and have agile and lean development methods to achieve competitive edge in the market (Lesser and Ban, 2016). The rapid transition from ideation to delivery is crucial for success in the marketplace.

In a global competitive landscape, it is difficult to have a sustainable competitive advantage. Organizations should develop a portfolio of innovation capabilities that can be harnessed to gain advantage (Holyzman, 2014). Innovation in organizations can be supported through strategic management of knowledge and competence (Inkinen et al., 2015). The compensation practices should be aligned with knowledge practices and information-technology practices.

7.5   ROLE OF HUMAN RESOURCES

The modern workplace is undergoing a fast change as a result of evolution in technology, demographic and social changes. The organizations have become more diverse in nature with the world becoming a global village. Gen Y workforce brings with it different values and work culture. The enforcement of regulations is stringent with public opinion and media playing a prominent role. The role of HR is also evolving simultaneously. It has come a long way from focus on just compliance adherence and data administration to being a strategic partner and fostering a right culture in the organization. The past decade has seen devolved management structures with more emphasis on first-line managerial leaders (Martins, 2015).

7.5.1   QUESTIONS FOR HR LEADERS

Longenecker and Fink (2015) provide a list of ten questions for the HR Leadership to assess their HR expertise. The questions form an easy checklist to review, reflect, and gauge the critical skills.

1. Do the members of your organization trust you?
2. Do you have complete knowledge of how the business makes money?
3. Are you up-to-date on the legal and compliance issues pertaining to your organization?
4. Are you a strategic thinker and operational executor?
5. Do you monitor and measure the key goal driven metrics?
6. Do you passionately create alliances with stakeholders to further business?

7. Do you help other managers to act as great HR leaders?
8. Is your influence and talent being used to create teams and solve issues?
9. Are you applying innovative HR practices to enhance the profitability of your organization?
10. Do you facilitate in timely information to employees to optimize performance?

The key areas identified can be worked upon to improve performance.

7.5.2 TALENT MANAGEMENT

The HR's function is to ensure that there are right people doing their job well across all departments and divisions. The higher the individual–organizational fit, the lower the turnover intention of the employees (Wei, 2015). As Davenport (2015) says it requires focus on the key areas of "job architecture, incumbent assessment, performance definition, recognition for success, and building leadership capacity." The adequate management and leadership capability at all levels ensure that the resources of the organization are efficiently utilized. It is leader that envisions the future direction that the organization is going to go in and makes sense of the same for others in the organization. The leader motivates the followers to achieve the goals. The managers deploy the resources and put the systems into practice to channelize the contribution of each individual towards the organizational goals. Davenport (2015) emphasizes the importance of "soft skills" which are hard to master and enable leaders to maximize the contribution of others.

The HR leaders can add value to the organization through the creation of innovative and technologically adept organization (Phillips and Phillips, 2016). They can support the other functional leaders in operations and make the organization competitive to take on the challenges posed by globalization. The ambidextrous leadership has been associated with team innovation (Zacher and Rosing, 2015) McKenzie and Aitken (2012) have developed a framework of 12 mutually supportive and complementary practices for leadership agility to create a conductive climate for knowledge sharing, engagement, collaboration, and learning. The framework can be used for self-assessment and develop an approach for change. It identifies the causes hampering learning in organizations and helps in overcoming the conflicts and tensions of organizational life.

The competency models are widely used to recruit, assess, organize, manage, and develop human capital in organizations. The methodology sets out the personality characteristics leaders should have and is closely linked to Goleman's (1998) concept of emotional intelligence comprising self-awareness, self-management, and social awareness. The models help in performance assessment, identifying high potential employees, and succession planning. The competency models have evolved from performance assessment, to alignment with strategic objectives, to facilitation of organizational change.

Information sharing and power sharing are leadership enablers whereas skill development act as leadership substitute (Doucet et al., 2015). Human resources play a mediation effect between leadership and total quality management for firm performance (Dubey et al., 2015). Mayo (2016) emphasizes that employee engagement and other measures of well-being should be clearly distinguished and that engagement should be measured frequently as it is linked to improved performance. Barr (2015) illustrated through a case study how an organization keeps its employees engaged through performance incentives, vision updates on a quarterly basis, and ingraining the values of the organization. These initiatives led to lower attrition rate and higher revenues, motivation, and employee satisfaction.

The HR management contributes towards organizational learning (Yazdani et al., 2016). The HR practices have a direct and indirect effect on employees' organizational commitment (Aladwan et al., 2015). The organizational climate plays a mediating role between the HR practices and employee outcomes (Cafferkey and Dundon, 2015). There is a solid link between employee engagement and improved performance (Mayo, 2016). Rae et al. (2015) found significant association between organizational affective commitment, employee performance process, training, and enhanced performance of the organization in environmental context. The researchers found a sequential chain between work practices, process improvement, and innovation process in an organization.

Odoardi et al. (2015) linked the managerial practices and leadership style to innovative work behavior of employees. The researchers found the participatory leadership style, information sharing, and teamwork led to group processes of team vision and support for innovation enhancing psychological empowerment of employees. The psychological empowerment of employees was positively associated with innovative behavior. Khoreva (2016) studied the association between leadership development practices and employee attitudes. The examination brought forth that affective commitment acts as a mediator between leadership development practices and commitment to accept increased performance demands and to build relevant competencies.

The human development practices lead to positive outcomes for the organization. The talent engagement initiatives and job-related training both have a positive impact on the quality of service provided by the employees of the organization (Wickramasinghe, 2015). Katou and Budhwar (2015) demonstrate through an empirical study on how improved training and development and increased compensation and incentives lead to higher productivity and competitiveness of the organization. Bish et al. (2015) found implementation of HR policies through divisional leaders leads to desired employee change-related outcomes. Renaud et al. (2015) conducted a longitudinal study of Canadian information technology sector employees. The researchers highlighted satisfaction with "respectful and stimulating work environment, training and development, innovative benefits, and incentive compensation" as crucial HR practices for reten-tion of experts and non-experts.

In knowledge intensive industries such as research and development, the innovation provides the basis for competitive advantage for the firms. The knowledge of the firm can get leaked through attrition of the key personnel working on projects and can jeopardize the successful comple-tion of projects. It is critical to protect the knowledge of the firm. Olander et al. (2015) examined the role that HR play in protection of knowledge of the organization. The researchers found that practices such as trust, moti-vation, commitment, and sense of responsibility are crucial to preserve the intellectual capital and knowledge of the organization.

From the perspective of a developing country, Kansal and Joshi (2015) found few HR-related disclosures in annual reports by top 82 Indian companies. The disclosed information mainly relates to training programs, safety awards/certifications, and the cordial relations between the manage-ment and workforce. There is a paucity of information on welfare schemes and benefits. Rao (2016) studied the HR management practices followed by the 25 companies identified as "best" in the year 2011. Out of these, 16 were Indian companies and 9 subsidiaries of multinational companies. The study shows light on progressive HR management practices grouped under seven themes namely: "elaborate staffing, investment in learning, work–life balance, egalitarian practices, developmental performance culture, generous benefits, and engagement initiatives." This shows that learning has become an important theme in organizations in the developing world too.

7.5.3   COLLABORATIVE CULTURE CREATION

Human resource development programs promote linkage between knowledge transfer and knowledge creation (Matsuo, 2015). The transfer of competencies provides the basis for creation of new knowledge. The on-the-job and off-the-job training programs facilitate the transfer of tacit and explicit knowledge. The interaction between the people, technology, and techniques helps in transfer of knowledge. The communities of practice also develop and implement HRD programs (Matsuo, 2015). The collaborative HR management practices create learning environment that fosters knowledge sharing behavior, which enhances individual as well as organizational capabilities (Iqbal et al., 2015).

The right organizational culture is important for employee engagement and improved firm performance. The competitive advantage of the organization can be sustained through nurturing a "learning by doing" type of culture. A culture, which is externally oriented, flexible, proactive and open, has a positive impact on the innovation performance (Laforet, 2016). On the other hand the culture with an inward focus impedes innovation performance. The winning culture transforms the customer promises into internal organizational actions. Ulrich and Brockbank (2016) found that the impact of organizational culture on firm's performance is 2–4 times that of individual employee's talent. The change in organizational culture which requires a change in attitude, beliefs, and values of its employees is time-consuming process. Coulson-Thomas (2015) found that behavior of key group of employees can be quickly changed independent of culture change through performance support so that the learning and routine working can simultaneously take place. This is an alternate route to change behavior. This raises the question whether behavior change through culture change is practical, desirable, affordable, and time efficient.

Secundo et al. (2016) have developed a novel framework for the management of intellectual capital of an organization through collective intelligence. The main components of the framework are goal of the organization (what); collective human capital (who); processes undertaken (how); and the motivations (why). The power of the intellectual capital is harnessed through its engagement with multiple stakeholders and value is created for the organization and society.

7.5.4 CHANGE MANAGEMENT

Human resource practices play a crucial role in employee perception and commitment during organizational change (Maheswari and Vohra, 2015). The practices of "culture, leadership, cross-functional integration, training, communication, and technology" have a positive impact on the employee perception and commitment. These practices should be assessed during change initiation, implementation, and consolidation in order to know about the employees' perceptions. The HR management function plays an important role of conserving HR during organizational acceleration. The continuous organizational change brings time pressures, work overload, occupational stress, and burnout threats. The holistic approach to HR ensures that individual resources of employees are sustained (Pluta and Rudawska, 2016). Sheehan et al. (2016) have assessed the value of HR during dynamic environments and conclude that role conflict and overload can limit the contribution of HR to strategic decision making. The HR systems flexibility in conjunction with traits such as self-efficacy and optimism in employees leads to innovative behavior (Wojtczuk-Turek and Tureck, 2015).

Zhang and Lv (2015) investigated the influence of contextual factor of environmental uncertainty on relationship between HR practices and new product development. The researchers found that HR exploitation influences new product development to a higher degree when environment uncertainty is low. Conversely, HR exploration influences new product development during periods of high environmental uncertainty.

7.5.5 MANAGING GAPS IN PERCEPTION

The different functions in an organization can have different perception pertaining to the implemented management systems in an organization.

Jaeger and Adair (2016) found that project managers and quality management representatives disagreed on their perception of the most important
benefit of the total quality management system. Quality was perceived to
be the most important benefit by project managers and productivity was
perceived to be the most important benefit by quality management representatives. The perception of various functions in an organization should
be aligned in order to gain optimum benefits. Huo et al. (2015) through
their interdisciplinary approach using data from ten countries found the
relevance of HR management practices for supply chain integration.

Cooper and Baird (2015) found a significant gap between the policies and
practices of flexible working arrangements in organizations. The procedure
and substantive outcomes of such requests were result of manager's personal
experience and levels of commitment. The conflicting organizational policies hinder the accessibility and effectiveness of the requests of employees.
There is an interaction between the formal and informal elements and role
and place of direct managers in implementation of policies.

There is a general lack of understanding of strategic thinking as a "recursive process of scanning, questioning, conceptualizing, and testing" among
HR leaders (Goldman et al., 2015). There is a poor alignment between the
design, delivery, and assessment of leadership development programs. The
satisfaction with the program is generally measured and little exploration of
change in behavior and its organizational impact is studied.

7.5.6   DIVERSITY MANAGEMENT

The active coordination results in team member's expertise being integrated in teamwork. In direct face-to-face interactions, it is easier to
coordinate. In dispersed teams, there are challenges of coordination and
transfer of understanding between team members and expertise coordination does not evolve automatically. The challenge can be overcome
through shared leadership where there is distribution of leadership. It can
be used as an initiating mechanism for expertise coordination in dispersed
teams (Muethel and Hoegi, 2016).

Reilly (2015) studied HR policies and practices in 70 organizations
and came to the conclusion that western organizations seem to be moving
toward a single global HR model which minimizes the cultural differences. On the other hand, the Asian organizations are more polycentric in

nature and retain distinct culture and diversity. The ownership structures and nature of business models may be responsible for this perspective. The standardization and consolidation of HR services reduces costs. The shift is toward centralization and imbibing one common philosophy in western organizations. This approach has the risk of organizations becoming ethno-centric promoting the home country perspective. This risk can be mitigated by having common people principles while retaining the distinct local prac-tices (Reilly, 2015). The standardization of the principles of human capital management is still in the nascent stage (Zheitoukhova, 2015). Divergent thing and brainstorming play an important role in organizational creativity, innovation, and change. The organizational culture can prove to be an impediment for brain storming and divergent thinking thus, limiting the potential for innovation and change (Kalargiros and Manning, 2015).

The devolved organizations have placed responsibility on the local level managerial decision-making. This reconfiguration in the roles has brought about more HR management responsibilities. Diversity management and leadership are the key issues for HR management (Martins, 2015).

Siakas and Siakas (2015) have developed an instrument to assess the cultural intelligence of partners in global transactions. The cultural issues when addressed with sensitivity lead to improved communication, mutual understanding, and effectiveness. The effective learning can take place through the close coordination of business and HR activities. The Gen Y has a different approach to work and careers. The changing demographic profiles and the emerging norms in societies have placed new challenges for the organizations. The organizational policies need to support the young professionals who are not constrained by the gender stereotypes of the previous generations (Clarke, 2015).

7.5.7   NEW INITIATIVES AT GENERAL ELECTRIC

General Electric (GE) with operations in over 140 countries with over 300,000 employees views inclusiveness and diversity as a cornerstone for productivity, creativity, innovation, and competitive advantage. The company has brought in new policies to attract and retain millennials. The company is doing away with the annual performance review system and replacing it with continuous feedback in the real time. The app "PD@GE" has been piloted on 3000 employees where constant feedback is provided by the managers. Other companies such as Cisco, Adobe, and LinkedIn have also introduced alternative ways to review the performance of employees. The role of managers is evolving as that of a coach. The managers guide the teams to achieve established goals. The company has launched its digital learning platforms and global learning centers for developing leadership skills. The company invests a billion dollars on learning and development each year. The company offers work flexibility to its employees to bring in better work–life balance. It has brought in greater parental leave benefits. The company provides 10 weeks of "GE Parental Leave." The new parents can take the help of the program named "GE Babies" where personalized guidance is provided by trained maternity nurses. Another novel program called "Moms on the Move" is being piloted that helps traveling nursing mothers in the United States to ship their milk back home for the baby. The new initiatives are poised to attract and retain the digital generation which craves for flexibility in work schedules, greater work–life balance, instant feedback, and fast career growth.

(Source: Jenkins, 2016)

7.6   PRACTICAL IMPLICATIONS

The global competitive landscape brings in new challenges on continuous basis. Small players can disrupt businesses and challenge the decades old rule of business. It is difficult for any organization to have a sustainable competitive advantage for a long duration of time. The new opportunities present new avenues for innovation to existing players to create and cater to the market needs in novel ways. The HR management plays an important role in creating and maintaining a conducive climate for knowledge creation, learning, and innovation in the organizations.

7.6.1   RESHAPE HR POLICIES

The HR practices should be adjusted to achieve the operational goals of the organization (Huo et al., 2015). Organizations need to develop and implement supportive policies to attract, maintain, and retain employees. The policies need to match up to the expectations of the changing demographic profiles and emerging social norms in the current setting. In knowledge creation, the workforce can demand a considerable amount of autonomy. The flexible working environment to engage talent should be adopted. The organization should reorient its policies and practices to provide a working environment that encourages cross-fertilization of ideas. Conflicting organizational policies need to be redeveloped for clear direction and implementation. The gap between the policies and practices needs to be bridged. The HR practices should encourage employees learning from their own and others professional experience to enhance contribution towards business goals. Recognition or other incentives need to be built in the systems so that ideas are disseminated (Downes and Marchant, 2016). The performance management system should evolve to include rewards for use of reflective work practices (Saunita et al., 2015). Multiple HR development programs should be integrated to link knowledge transfer and knowledge creation by a combination of on-the-job training, off-the-job training, and continuous improvement programs (Matsuo, 2015).

7.6.2   USE OF AUTOMATION AND TALENT ANALYTICS

The use of big data and talent analytics has opened doors for real-time analysis of performance. The routine tasks such as payroll and induction should be automated in-house or outsourced. The efficient and productive organization should enhance capabilities in specialist roles to be a differentiator in the market. The HR function should take a proactive approach in enhancement and development of talent in organization. The use of talent analytics should be made to recruit and retain talent. The skills gap analysis should be carried out for design of effective training and development programs. The analytics also help in gauging the mood in the organization on a real-time basis. The text analysis and regular sentiment surveys are windows to the minds of the employees. The sentiments of the workforce should be respected and efforts made to keep it engaged and committed. The talented and engaged human capital leads to innovation in organizations.

7.6.3   ENHANCEMENT OF CROSS-CULTURAL INTELLIGENCE

The transnational organizations have boosted the global trade. The companies now have access to global talent pools. The organizations need to manage the cultural distances within its workforce and ingrain the corporate culture in all its employees. The harmonious relations between the global partners are based on cultural intelligence. Organizations having global transactions need to assess, develop, and monitor the cultural intelligence of employees in order to have harmonious and mutually beneficial relations. Cross-cultural intelligence needs to be developed in the employees to enhance performance.

7.6.4   APPRECIATION AND AWARENESS OF GENDER, AGE, AND ETHNIC DIVERSITY

The diverse demographic profile of employees in various nations poses a challenge for the organizations. The organizations need to address gender and ethnic diversity in workforce. The organizations of the future also face issues of age diversity. The older employees in developed countries and the digital generation need to work in tandem for smooth operations. The conflict between the work behaviors and values need to be minimized. The continued development of aging workforce should be taken into consideration while formulating policies (Warhurst and Black, 2015). Employees need to be trained and aligned with the organizational values and purpose. The flat organization structures have placed greater responsibility with the frontline managers. The diversity in the organization needs to feed innovation and organizational well-being. A holistic strategic approach needs to be developed to cater to diversity and enhanced role of frontline managers (Martins, 2015).

7.6.5   BUILDING AGILE AND FLEXIBLE ORGANIZATION

In the dynamic environment, the organization for the future need be agile and flexible. It should be able to quickly change its form and structure to align with rapid expansion or rationalization according to business needs. The design thinking needs to be incorporated in the organization. The project teams should be able to quickly align and work together according to the objectives and dismantle when the goal has been achieved. The contractual employment provides a solution to quickly expand capacity and take care of sudden demands at a reasonable cost. The flexibility of the organization is increased but the short-term workforce needs to be aligned with the organizational goals to maintain productivity. The transfer of knowledge from contractual employees to permanent employees is another challenge.

7.6.6   USE OF COMPETENCY MODELS

The competency models should ensure a link to the business goals and organizational strategies. The future job requirements should be taken into consideration along with the organizational context and goals while developing competency models. The communications to employees should make use of heuristics and diagrams. The competencies should be aligned with recruitment, performance assessment, promotion, compensation, and development.

7.6.7   ORGANIZATIONAL COMMITMENT FOR KNOWLEDGE SHARING

The employees of the organizations may or may not be willing to share knowledge. The willingness of the employee to give and receive knowledge increases the overall pool of organizational knowledge. The willingness or reluctance of the employees should be studied. The motivations behind knowledge sharing behaviors should be understood. The organizational commitment is linked to knowledge sharing behavior and attitude. Organizations should deploy HR practices that attract and retain the talented employees (Wei, 2015). The policies which ensure good match between employee and organization lead to retention of outstanding employees. Policies should be framed to keep the workforce committed and engaged. The knowledge sharing enablers and processes should be put in place to sustain innovation.

7.6.8   SHARED LEADERSHIP FOR DISPERSED TEAMS

The challenges of managing dispersed teams in multiple countries and different time-zones should be overcome through effective coordination and control. Shared leadership should be used as a coordination mechanism for transfer of knowledge among dispersed team members (Muethel and Hoegi, 2016).

7.6.9   EFFECTIVE TEAMS FOR INNOVATION

The innovative behavior in employees can be mobilized through group processes such as team vision and support for innovation. The requisite behavior of employees should be nurtured through development of team work, effective communication system, and participatory skills of managers (Odoardi et al., 2015). The framework developed by McKenzie and Aitken (2012) for leadership agility practices can be used by the leaders for self-assessment. The approach to change and learning can be altered based on the results of assessment. The common causes of tensions and conflicts in the organizational life can be overcome to steer learning. The leadership development programs design, delivery methods, and impact assessment should be properly aligned.

7.6.10   WINNING ORGANIZATIONAL CULTURE

The role of organizational culture should be understood and utilized to enhance business results. The impact of culture is 2–4 times greater than individual talent (Ulrich and Brockbank, 2016). The culture should take into consideration diverse viewpoints. Divergent thinking should be encouraged. Brainstorming and divergent thinking lead to innovation. The organizational culture should be molded in such a way that change and innovation ability of the organization is enhanced. The collaboration and co-creation ecosystem should be developed. Conducive organizational climate needs to be created to foster innovation in order to build resilient organizations. The inculcation of winning organizational culture is important for achievement and sustainability of competitive advantage.

7.6.11   INCORPORATION OF STRATEGIC THINKING

Organizations for the future need to develop employees capable of conceptual, holistic, and opportunistic thinking through the recursive process of scanning, questioning, conceptualizing, and testing to derive novel solutions organizations (Goldman et al., 2015). Strategic thinking is important for employees working at all levels of the organization. It is especially so for those organizations that promote from within. Strategic thinking needs to be incorporated in the talent management systems in the organizations. The leadership programs in organizations serve the purpose of conceptual and theoretical understanding, skill building, feedback, and personal growth. The knowledge and skills can be enhanced and behavior can be changed during formal training. The development can take place through mentoring, outdoor programs, job rotation, special assignments, and assessment centers. Interactive computer programs, instructional videos and source material can be self-help guides. Conceptual, procedural, and factual knowledge is enhanced through each experience. Leadership and strategy scholars and practitioners should work together to design effective programs incorporating content, practice, feedback, and reflection (Goldman et al., 2015). Organizations should train ambidextrous leaders as such leadership behaviors predict innovation (Zacher and Rosing, 2015).

The HR function should be seen as a trusted ally, advisor, and coach in the organization (Davenport, 2015). The organizations of the future require employees with an entrepreneurial attitude who view change as an opportunity rather than a challenge. Organizations should promote knowledge sharing culture to enhance innovative capability. The organizations with knowledge management capability are able to efficiently use resources and build competitive advantage.

7.7 CONCLUSION

The role of HR is paramount in an organization. Development of competent and capable executives is continuous challenge for many organizations in the current volatile business environment (O'Neill and Bent, 2015). The HR practices need to meet the challenge of the times and inculcate dynamic capability of learning. The innovative culture needs to be embedded deep in the organization across all divisions and functions. The organization itself needs to be able to change its form and structure in a short duration of time. The organizations should be able to access requisite resources easily and promptly to harness advantages offered by the marketplace. The project teams should be able to organize resources from across divisions and functions in a short duration of time in an integrated manner. The same swiftness and agility should be shown during unbundling process, and the teams should disappear with ease when no longer required.

The organizations have to transform to take advantage of the new opportunities offered and meet up to the challenges posed by the environment. The ingrained ways of work have to be altered to stand up to new challenges. The launch of new products and services requires development, transformation and deployment of knowledge assets. Leadership and organizational practices that support efficient and effective management of knowledge should be pursued to enhance organization's performance (Inkinen, 2016). The knowledge of multiple individuals needs to be integrated in the organizational systems and processes. The collaborative process of knowledge creation needs to be encouraged in organizations. The only course that would lead to sustainable growth is to make the organization capable of learning. To do so the culture of experimentation needs to be inculcated. The employees should not be afraid of trying out new things at work. The open culture should encourage team work and formation of support groups. The failures should not be punished and should become stepping stones to future success. The organizational learning capability needs to be enhanced to make the organization more innovative.

KEYWORDS

- **knowledge management**
- **Innovation**
- **learning organization**
- **future HR practices**
- **strategic leadership**
- **design thinking**
- **organizational change**

REFERENCES

Aladwan, K.; Bhanugopan, R.; D'Netto, B. The Effects of Human Resource Management Practices on Employees' Organizational Commitment. *Int. J. Organ. Anal.* **2015,** *23*(3), 472–492.

Amankwah-Amoah, J.; Zhang, H. Tales From the Grave: What Can we Learn From Failed International Companies? *Foresight* **2015**, *17*(5), 528–541.

Barr, K. Engaged Employees Shimmer for Oasis HR: Fast-Growing Agency Wins Prize for Three-Prong approach. *Hum. Resour. Manage. Int. Dig.* **2015**, *23*(3), 5–7.

Bish, A; Newton, C.; Johnson, K. Leader, Vision and Diffusion of HR Policy During Change, *J. Org. Change Manage.* **2015**, *28*(4), 529–545.

Brink, T.; Madsen, S. O. Entrepreneurial Learning Requires Action on the Meaning Generated. *Int. J. Entrepreneurial Beha. Res.* **2015**, *21*(5), 650–672.

Cafferkey, K.; Dundon, T. Explaining the Black Box: HPWS and Organisational Climate. *Pers. Rev.* **2015**, *44*(5), 666–688.

Campbell, P.; Evans P. Reciprocal Benefits, Legacy and Risk: Applying Ellinger and Bostrom's Model of Line Manager Role Identity as Facilitators of Learning. *Eur. J. Train. Dev.* **2016**, *40*(2), 74–89.

Clarke, M. Dual careers: the new norm of Gen Y professionals? *Career Dev. Int.* **2015**, *20*(6), 562–582.

Cooper, R.; Baird, M. Bringing the "Right to Request" Flexible Working Hours Arrangements to Life from Policies to Practices. *Employee Relat.* **2015**, *37*(5), 568–581.

Corfield A.; Paton, R. Investigating Knowledge Management: can KM Really Change Organisational Culture? *J. Knowl. Manage.* **2016**, *20*(1), 88–103.

Coulson-Thomas, C. Learning and behaviour addressing the culture change conundrum: part two. *Ind. Commer. Train.* **2015**, *47*(4), 182–189.

Davenport, T. O. How HR Plays Its Role in Leadership Development. *Strategic HR Rev.* **2015**, *14*(3), 89–93.

Doos, M.; Johansson, P.; Withelmson, L. Beyond Being Present: Learning-Oriented Leadership in the Daily Work of Middle Managers. *J. Workplace Learn.* **2015**, *27*(6), 408–425.

Doucet, O.; Lapaime, M.; Simard, G.; Tremblay, M. High Involvement Management Practices as Leadership Enhancers. *Int. J. Manpower* **2015**, *36*(7), 1059–1071.

Downes, T.; Marchant, T. The Extent and Effectiveness of Knowledge Management in Australian Community Service Organisations. *J. Knowl. Manage.* **2016**, *20*(1), 49–68.

Dubey, R.; Singh, T.; Ali, S. S. The Mediating Effect of Human Resource on Total Quality Management Implementation: An Empirical Study on SMEs in Manufacturing Sectors. *Benchmarking Int. J.* **2015**, *22*(7), 1463–1480.

Ekanem, I. Entrepreneurial Learning: Gender Differences. *Int. J. Entrepreneurial Beha. Res.* **2015**, *21*(4), 557–577.

Engestorm, Y. *Learning by Expanding: An Activity-Theoretical Approach to Developmental Research*, 2nd ed.; Cambridge University Press: New York, 2015.

Ernst & Young. *Imagining the Digital future: How digital themes are transforming companies across industries*. 2015.

Goldman, E. F.; Scott, A. R.; Follman, A. M. Organisational practices to develop strategic thinking. *J. Strategy Manage.* **2015**, *8*(2), 155–175.

Goleman, D. What Makes a Leader. *Harv. Bus. Rev.* **1998**, *Nov-Dec.*, 92–102.

Haase, H.; France, M.; Felix, M. Organisational Learning and Intrapreneurship: Evidence of Interrelated Concepts. *Leadership Org. Dev. J.* **2015**, *36*(8), 906–926.

Helyer, R. Learning Through Reflection: the Critical Role of Work-Based Learning (WBL). *J. Work-Appl. Manage.* **2015**, *7*(1), 15–27.

Holyzman, T. A Strategy of Innovation Through the Development of a Portfolio of Innovation Capabilities. *J. Manage. Dev.* **2014,** *33*(1), 24–31.

Huo B.; Han, Z.; Chen, H.; Zhao, X. The Effect of High-Involvement Human Resource Management Practices on Supply Chain Integration. *Int. J. Phys. Distrib. Logistics Manage.* **2015,** *45*(8), 716–746.

Impedovo, M. A.; Manuti, A. Boundary Objects as Connectors Between Communities of Practices in the Organizational Context. *Dev. Learn. Organ. Int. J.* **2016,** *30*(2), 7–10.

Inkinen, H. Review of Empirical Research on Knowledge Management Practices and Firm Performance. *J. Knowl. Manage.* **2016***, 20*(2), 230–257.

Inkinen, H. T.; Kianto, A.; Vnhala, M. Knowledge Management Practices and Innovation Performance in Finland. *Balt. J. Manage.* **2015,** *10*(4), 432–455.

Iqbal, S.; Toulson, P.; Tweed, D. Employees as Performers in Knowledge Intensive Firms: Role of Knowledge Sharing. *Int. J. Manpower* **2015,** *36*(7), 1071–1094.

Jaeger, M.; Adair, D. Perception of TQM Benefits, Practices and Obstacles: The Case of Project Managers and Quality Management Representatives in Kuwait. *TQM J* **2016,** *28*(2), 317–336.

Jenkins, R. 5 New GE Initiatives that will Attract and Retain Millenials. 2016. http://www.inc.com/ryan-jenkins/5-new-ge-initiatives-that-will-attract-and-retain-millennials.html (accessed Sep 29, 2016).

Kalargiros, E. M.; Manning, M. R. Divergent Thinking and Brainstorming in Perspective: Implications for Organizational Change and Innovation in *Research in Organ. Change and Development*; Shani, A. B., Noumair D. A., Eds.; Emerald Group Publishing Limited: Bingley, West Yorkshire, 2015; Vol. 23, pp 293–327.

Kansal, M.; Joshi, M. Reporting Human Resources in Annual Reports: An Empirical Evidence From Top Indian Companies. *Asian Rev. Account.* **2015,** *23*(3), 256–274.

Katou, A. A.; Budhwar, P. Human Resource Development and Organisational Productivity: A Systems Approach to Empirical Analysis. *J. Organ. Eff. People Perform.* **2015,** *2*(3), 244–266.

Khoreva, V. Leadership Development Practices as Drivers of Employee Attitudes. *J. Managerial Psychol.* **2016,** *31*(2), 537–551.

Knight, E.; Harvay, W. Managing Exploration and Exploitation Paradoxes in Creative Organisations. *Manage. Decis.* **2015,** *53*(4), 809–827.

Laforet, S. Effects of Organisational Culture on Organisational Innovation Performance in Family Firms. *J. Small Bus. Enterp. Dev.* **2016,** *23*(2), 379–407.

Lancaster, S.; Milia, L. D. Developing a Supportive Learning Environment in Newly Formed Organisation. *J. Workplace Learn.* **2015,** *27*(6), 442–456.

Lesser, E.; Ban, L. How Leading Companies Practice Software Development and Delivery to Achieve Competitive Edge. *Strategy Leadership* **2016,** *44*(1), 41–47.

Limwichitr, S.; Broady-Preston, J.; Ellis, D. A Discussion of Problems in Implementing Organisational Cultural Change: Developing a Learning Organisation in University Libraries. *Library Rev.* **2015***, 64*(6/7), 480–488.

Longenecker, C.; Fink, L. Ten Questions That Make a Difference for HR Leadership: The Distilled Wisdom of Two Award Facilitators. *Hum. Resour. Manage. Int. Dig.* **2015,** 23(3), 20–22.

Mafabi, S.; Munene, J. C.; Ahiauzu, A. Creative Climate and Organisational Resilience: the Mediating Role of Innovation. *Int. J. Organ. Analysis*, **2015,** *23*(4), 564–587.

Maheswari, S.; Vohra, V. Identifying Critical HR Practices Impacting Employee Perception and Commitment During Organizational Change. *J. Organ. Change Manage.* **2015,** *28*(5), 872–894.

Martins, L. HR Leaders Hold the Key to Effective Diversity Management—as More and More Important Decisions are Taken at Local Level. *Hum. Resour. Manage. Int. Dig.* **2015,** 23(5), 49–53.

Matsuo, M. Human Resource Development Programs for Knowledge Transfer and Creation: The Case of the Toyota Technical Development Corporation. *J. Knowl. Manage.* **2015,** *19*(6), 1186–1203.

Mayo, A. The Measurement of Engagement. *Strategic HR Rev.* **2016,** *15*(2), 83–89.

McKenzie, J.; Aitken, P. Learning to Lead knowledgeable Organization: Developing Leadership Agility. *Strategic HR Rev.* **2012,** *11*(6), 329–334.

Miller, F.Q. Experiencing Information Use for Early Academics Learning: a Knowledge Ecosystem Model. *J. Documentation,* **2015,** *71*(6), 1228–1249.

Muethel, M.; Hoegi, M. Expertise Coordination over Distance: Shared Leadership in Dispersed New Product Development Teams, in *Leadership Lessons from Compelling Contexts;* Peus, C., Braun, S., Schvns, B. Eds.; Emerald Publishing Group Limited: Bingley, West Yorkshire, 2016; Monographs in Leadership and Management Vol. 8, pp 327–348.

Odoardi, C. Montani, F. Boudnas, J. and Battisteli, A. Linking Managerial Practices and Leadership Style to Innovative Work Behavior: The Role of Group and Psychological Processes. *Leadership Org. Dev. J.* **2015,** 36, 5, 545-569.

Olander, H.; Hurmelinna-Laukkanen, P.; Heilmann, P. Human Resources—Strength and Weakness in Protection of Intellectual Capital. *J. Intellect. Cap.* **2015,** *16*(4), 742–762.

Ollila, S.; Ystrom, A. Authoring Open Innovation: The Managerial Practices of an Open Innovation Director. *Research in Organ. Change and Development;* Shani, A. B., Noumair, D. A. Eds.; Emerald Group Publishing Limited, 2015; Vol. 23, pp 253–291.

O'Neill, A. B.; Bent, R. The Advantages of a Transorganisational Approach for Developing Senior Executives. *J. Manage. Dev.* **2015,** *34*(5), 621–631.

Paraponaris, C.; Sigal, M. From Knowledge to Knowing, From Boundaries to Boundary Construction. *J. Knowl. Manage.* **2015,** *19*(5), 881–899.

Phillips, J.J.; Phillips, P. How can HR have an Impact on Non-Traditional Areas. *Strategic HR Rev.* **2016,** *15*(1), 5–13.

Pluta, A.; Rudawska, A. Holistic Approach to Human Resources and Organizational Acceleration. *J. Organ. Change Manage.* **2016,** *29*(2), 293–309.

Rae, K.; Sands, J.; Gadenne, D. L. Associations Between Organisations Motivated Workforce and Environmental Performance. *J. Accounting Org. Change* **2015,** *11*(3), 384–405.

Rao, P. Investment and Collaboration: the Indian Model for "best" HRM Practices. *J. Asia Bus. Stud.* **2016,** *10*(2), 125–147.

Reilly, P. Managing Across Borders and Cultures. *Strategic HR Rev.* **2015,** *14*(1/2), 36–41.

Renaud, S.; Morin, L.; Saulquin, J.; Abrahan, J. What are the Best HRM Practices for Retaining Experts? A Longitudinal Study in the Canadian Information Technology Sector. *Int. J. Manpower* **2015,** *36*(3), 416–432.

Saunita, M.; Tillamaki, K.; Ukko, J. Managing Performance and Learning Through Reflective Practices. *J. Organ. Eff. People Perform.* **2015,** *2*(4), 370–390.

Secundo, G.; Dumay, J. Schiuma, G.; Passiante, G. Managing Intellectual Capital Through a Collective Intelligence Approach: An Integrated Framework for Universities. *J. Intellect. Cap.* **2016**, *17*(2), 298–319.

Senge, P. *The Fifth Discipline: The Art and Practice of the Learning Organisation*, Century Business: London, 1990

Sheehan, C.; Cieri, H. D.; Cooper, B.; Shea, T. Strategic Implications of HR Role Management in a Dynamic Environment. *Pers. Rev.* **2016**, *45*(2), 353–373.

Siakas, K.; Siakas, D. Cultural and Organisational Diversity: Evaluation (CODE): A Tool for Improving Global Transactions. *Strategic Outsourcing Int. J.* **2015**, *8*(2/3), 206–228.

Smith, S. and Martin, J. Practitioner Capability: Supporting Critical Reflection during Work based Placement—A Pilot Study. *Higher Educ. Skills Work-Based Learn.* **2014**, *4*(3), 284–300.

Staldier, R.; Eullagar, S. Appreciating Formal and Informal Knowledge Transfer Practices Within Creative Festival Organizations. *J. Knowl. Manage.* **2016**, *20*(1), 146–161.

Sutton, C. The Human Sigma Approach to Business Improvement in Tourism SMEs. *J. Small Bus. Enterp. Dev.***2015**, *22*(2), 302–319.

Tam, S. and Gray, D. E. Organisational Learning and the Organisational Lifecycle: The Differential Aspects of an Integrated Relationship in SMEs. *Eur. J. Train. Dev.* **2016**, *40*(1), 2–20.

The Economic Intelligence Unit Limited (2015*) Engaging and Integrating a Global Workforce: Global Trends Impacting the Future of HR Management*, SHRM Foundation.

Ulrich, D. and Broackbank, W. Creating winning culture: next step for leading HR professionals. *Strategic HR Rev.* **2016**, 15(2), 51–56.

Warhurst, R. P.; Black, K. E. It's Never Too Late to Learn. *J. Workplace Learn.* **2015**, *27*(8), 457–472.

Wei, Y. Do Employees High in General Human Capital Tend to have a Higher Turnover Intention? The Moderating Role of High-Performance HR Practices and P-O Fit. *Personnel Rev.* **2015**, **44**(5), 739–756.

Wickramasinghe, V. Effects of Human Resource Development Practices on Service Quality of Services Offshore Outsourcing Firms. *Int. J. Qua. Reliab. Manage.* **2015**, *32*(7), 703–717.

Wojtczuk-Turek, A.; Tureck, D. Innovative Behaviour in Workplace: The Role of HR Flexibility, Individual Flexibility and Psychological Capital: the Case of Poland. *Eur. J. Innovation Manage.* **2015**, *18*(3), 397–419.

World Trade Organisation. *World Trade Statistical Review.* 2016. https://www.wto.org/english/res_e/statis_e/wts2016_e/wts2016_e.pdf (accessed Sep 27, 2016)

Yazdani, B.; Altafar, A.; Shanin, A.; Kheradmandnia, M. The Impact of TQM Practices on Organizational Learning Case Study: Automobile Part Manufacturing and Suppliers of Iran. *Int. J. Qua. Reliab. Manage.* **2016**, *33*(5), 574–596.

Zacher, H.; Rosing, K. Ambidextrous Leadership and Team Innovation. *Leadership Org. Dev. J.* **2015**, *36*(1), 54–68.

Zhang, H.; Lv, S. Effect of HR Practice on NPD Performance: The Moderating Role of Environmental Uncertainty. *Nankai Bus. Rev. Int.* **2015**, *6*(3), 256–280.

Zheitoukhova, K. New Ways of Working: What is the Real Impact on the HR Profession? *Strategic HR Rev.* **2015**, *14*(5), 163–167.

CHAPTER 8

CREATIVITY AND INNOVATION: THE FUTURE OF ORGANIZATIONS

NANDINI SRIVASTAVA

Faculty of Management, Manav Rachna International Institute of research and studies, Faridabad, India;
Email: nandini.fms@mriu.edu.in

CONTENTS

ABSTRACT

Creativity and innovation will be an indispensable part of future organizations due to every changing business environment and impact of globalization. This chapter tries to explore the impact of creativity and innovation on human resource (HR) practices in the scenario of globalization and how companies can be benefited and react to the need of the sustainable HR practices. The employee of the present age does not only look forward of the challenging and paying job but also looks for his holistic development which can be only once the organization are creative and innovative. The impact and need of creativity and innovation in organizations and how HR practices can be formed around the concepts is discussed.

The market potential of innovations seems too small relative to the size of the existing business. Big companies miss out on many important innovations because the potential is often inaccurately viewed as too small. Human resource professional should provide freedom and support to the new innovative ideas by keeping the strategies and long-term objectives of the organization in mind. Manager should give specific directions about ends, but leave the means to employees.

8.1   INTRODUCTION

The importance for the organizations to be more competitive has gained high interest of researchers to understand and inculcate creativity in organizations. Researchers have been able to establish the relationship between individual and organizational creativity and innovation (Amabile, 1996; Mumford et al., 2002).

The demonstration of the relationship between individual, team and organizational aspects of creativity (Woodman et al., 1993) has also been researched. To encourage the creativity organizations need to create a climate that supports and enables the creative thinking of employees (Amabile, 1988). Andriopoulos (2001) identified five major organizational dimensions under which characteristics and behaviors are placed that enhance or inhibit creativity in the work environment. Those dimensions are organizational climate, culture, structure and systems, leadership style, resources and skills.

The Encyclopedia Britannica defines creativity as "the ability to produce something new through imaginative skill, whether a new solution to a problem, a new method or device, or a new artistic object or form." At an individual level, Amabile's (1997) suggests that individual creativity essentially requires expertise (knowledge, proficiencies, and abilities of people to make creative contributions to their fields), creative thinking skills (cognitive styles, cognitive strategies, and personality variables), and intrinsic task motivation (the desire to work on something because it is interesting, involving, challenging, and rewarding). The study confirms that as the levels move higher on each of these three components, the higher and better the creativity.

The Wikipedia definition of creativity—"the ability of a person to be creative, participate in creating or be useful in a creative network of other people"—is a useful one that is both simple and broad enough to encompass both individuals and organizations.

Recent views on organizational creativity appear to be focusing largely on outcomes or creative products—goods and services. A creative product has been defined as one that is both novel and original and potentially useful or appropriate to the organization (Amabile, 1996, Mumford and Gustafson, 1998).

Various factors contribute to the generation of creative products both at the individual and organizational levels (Mumford and Gustafson, 1998). In organizations including businesses, creativity is the process through which new ideas that make innovation possible are developed (Paulus and Nijstad, 2003).

Innovation, in today's scenario is often used interchangeably with creativity. A basic definition of innovation from an organizational perspective is given by Luecke and Katz (2003), they defined the innovation as the introduction of a new thing or method. Innovation is the embodiment, combination, or synthesis of knowledge in original, relevant, valued new products, processes, or services". Innovation, it seems, typically involves creativity, but they are not identical. In an organization, it has been observed that for innovation to occur, something more than the generation of a creative idea or insight is required. The insight must be put into action to make a genuine difference.

8.2   PRESENT STAGE OF KNOWLEDGE

8.2.1   CREATIVITY

According to Wikipedia creativity can be defined as "the ability of a person to be creative, participate in creating or be useful in a creative network of other people." Woodman (1995) defines the organizational creativity as "the creation of a valuable, useful new product, service, idea, procedure or process by individuals working within a complex social organization." Creativity is the process by which individuals or small groups produce novel and useful ideas. Creativity in organizations is based on three fundamental components:

- Domain-relevant skills (basic knowledge to perform the task)
- Creativity-relevant skills (special abilities to generate new ideas)
- Intrinsic task motivation (willingness to perform creative acts)

Creativity in organizations may be promoted by training people to be creative, by encouraging openness to new ideas (e.g., "thinking outside the box"), by taking the time to understand the problem at hand, and by developing divergent thinking. It also may be accomplished by developing creative work environments. These are ones in which autonomy is provided, ideas are permitted to cross-pollinate, jobs are made intrinsically interesting, creative goals are set, creativity is supported within the organization, people have fun, and diversity is promoted.

8.2.2   INNOVATION

Innovation refers to the implementation of creative ideas within organizations. Innovation is composed of three components that are analogous to the three components of creativity. These are motivation to innovate, resources to innovate, and innovation management. These components are used in a process that generally proceeds in five stages: setting the agenda, setting the stage, producing the ideas, testing and implementing the ideas, and assessing the outcome.

8.2.3 *PRESENT STAGE OF HR IN INDIAN COMPANIES*

In the changing dynamics of Indian Market, companies are under tremendous pressure to become employer of choice among the competitors. There are many factors which are linked to the companies to become EOC for the workforce. The factors range from external to internal factors

1. External factors (Market)
   a. Generic orientation of market
   b. Intensively competitive market place
   c. Product portfolio strategies—mass market vs. niche market, and so forth.
2. Internal factors (Organizational)
   a. Knowledge workers, skills set mapping
   b. Sales force effectiveness
   c. Motivating field force for improving productivity
   d. Attracting and retaining talent with productive contribution
   e. Need for creative and innovative HR policies, and so forth.

8.3 IMPACT OF CREATIVITY AND INNOVATION ON HR PRACTICES

Human resource departments (HRDs) of organizations have been facing problems in terms of retaining the talents and striving hard to acquire some adoptable methods which could help them to be distinctive in the market or with competitors. From organization's point of view, HR has limited monetary and nonmonetary resources to keep the field force motivated for improving productivity year after year.

To understand the concept we should be able to understand the following points

1. Understand creativity as an effective management tool
2. Various creative HR approaches in organizations
3. Effectiveness of these approaches
4. To analyze how creativity can help in improving the employee productivity

5. Understanding focus need area for near future
6. Recommending/Discussing some creative ideas

8.4 UNDERSTANDING KEY ELEMENTS

Organizations are becoming large and complex with progressive indus-
trialization and advent of new technologies. Over the years government
interventions in regulating organizational purpose and performance has
increased. Organizations combine science and people—technology and
humanity. Organization's success and future depend upon the effective
utilization of the most important source—people working for them which
is very complex in itself to understand. Human behavior in organizations
is rather unpredictable because it arises from people's deep-seated needs
and value systems. There are no simple cookbook formulas for working
with people. There is no idealistic solution to organizational problems.
The goals are challenging and worthwhile. Organizations are nothing
without HR and to make their effective utilization, the organizations need
to be creative with respect to the company policies and satisfying human
needs. In addition to that for business organizations, creative ideas must
have utility. They must constitute an appropriate response to fill a gap in
the production, marketing, or administrative processes of the organization.
Thus, individual creativity is concerned with the generation of ideas while
team and organizational creativity is concerned with both the generation
of ideas and the implementation of these ideas.

With this topic we would like to evaluate and analyze the current status
and usage of creativity in industry which makes them employer of choice
on one hand and getting optimum productivity on the other hand.

8.5 CREATIVITY IN ORGANIZATIONS

1. Creativity in individuals and teams involves three basic components as follows: domain-relevant skills, creativity-relevant skills,
   and intrinsic task motivation.
   a. Domain-relevant skills

- Virtually any task requires certain talents, knowledge, or skills. Those skills and abilities we already have constitute the raw materials of creativity.
  b. Creativity-relevant skills
    - Beyond the basic skills, being creative also requires additional skills and, special abilities that help people to approach what they do in novel ways.
  c. Intrinsic task motivation
    - Intrinsic task motivation focuses on what people are willing to do.
    - To be creative, a person must be willing to perform the task in question.

8.6 TRAINING PEOPLE TO BE CREATIVE

- Anyone can develop the skills to be creative.
- Becoming more creative requires allowing oneself to be open to new ideas. Some companies send their employees on "thinking expeditions." These trips are designed to put people in challenging situations in an effort to help them think differently and become more creatively.
- Take time to understand the problem. Meaningful ideas rarely come to those who don't fully understand the problem at hand.
- Teaching people various tactics for divergent thinking allows problems to incubate, setting the stage for new ideas to develop. One popular way of developing divergent thinking is known as "morphology."
- A morphological analysis of a problem involves identifying the basic elements of a problem and systematically combining them in different ways.

0.7 DEVELOPING CREATIVE WORK ENVIRONMENTS

There are also steps the organization can take to bring out people's creativity. They include:

a. Providing autonomy: People are more creative when they have the freedom to control their own behavior and are empowered to make decisions.
b. Allow ideas to cross-pollinate: People who work on several projects pick up ideas for others that may have applicability elsewhere.
c. Make jobs intrinsically interesting: Creativity is enhanced the more interesting the job—turn work into play by making it interesting.
d. Set creative goals: Edison set a goal of a minor invention every 10 days and a major invention every 6 months and was an extremely productive inventor. Setting goals enhances creativity.
e. Support creativity at high organizational levels: Bosses must encourage employees to take risks.
f. Have fun: Providing employees the opportunity to have fun on the job is a powerful incentive.
g. Promote diversity: People from different backgrounds are bound to think differently about situations they face.

8.8   THE PROCESS OF INNOVATION

8.8.1   COMPONENTS OF INNOVATION: BASIC BUILDING BLOCKS

1. *Motivation to innovate*: Just as individual creativity requires that people be motivated to do what it takes to be creative, organizational innovation requires culture that encourage innovation.
2. *Resources to innovate*: Just as people must have certain basic skills to be creative, organizations must possess certain basic resources to make innovation possible.
3. *Innovation management*: Just as individuals must home special skills to be creative, organizations must develop special ways of managing people to encourage innovation, skills in innovation management.

4. Managers promote innovation when they show balance with respect to three key matters:
   a. Organizational innovation is promoted when goals are carefully linked with the corporate mission but not too specific.
   b. Reward systems should generously and fairly recognize one's contributions, but not connected to every move.
   c. Innovation management requires carefully balancing the time pressures under which employees are placed.

8.9   KEY CHALLENGES FACED BY HRD

Today's workforce is changing its face every day with the modernization, technology and globalization. Organizations need to be proactive on certain issues to achieve success and be an employer of choice for the workforce. Few such factors are analyzed here with the same mindset.

8.9.1   CHANGING DEMOGRAPHICS

First and foremost change is the demographic change in the workforce. This has really altered the composition of workforce significantly so the manner in which it needs to be managed should definitely change significantly. The educated, diverse, and empowered workforce will present challenges to managers in the current and future scenario. The highly skilled and educated employees in the knowledge oriented industries (for example Pharma, R&D, IT and IT-enabled services) are in a unique position—supply and demand creates many employment opportunities. Attracting the right required talent, recruiting and retaining these highly volatile individuals pose problems for managers in these organizations. Globalization will see more and more employees traveling abroad for their organizations, sharing acquired skill set across boundaries and influencing the host culture.

All over the world the role of women has changed drastically over the years as integral part of workforce, giving rise to newer issues such as women empowerment, discrimination issues, development plans as per family life cycle etc.

8.9.2   WORKFORCE DIVERSITY

Managing diversity is one of the hot issues even for relatively better-managed organizations in the 21st century. Organizations are perhaps in learning phase in terms of coping with a diverse workforce and impacting factors. Diversity in terms of cultural backgrounds, varied value systems, demographic compositions, geographic decentralization are a few examples of this felt diversification. Another concern is that of managing contract employees or Marketing JVs.

8.9.3   CHANGING ROLES

In the recent past, organizations have witnessed a major shift in roles played by various stakeholders in overall process as compared to ones in traditional approach. In many cases, these roles are modified and redefined for a better-managed scenario. Following are a few examples of changing roles:

1. Demanding jobs with in-depth knowledge
2. Concept selling
3. Performance orientation
4. Use of technology

These trends indicate the changing roles of sales force and managers, with empowerment of the sales force becoming a reality with added pinch of pressure to perform.

The field of HR is very vast and challenging. With increasing diversity in work force the challenges faced by managers are never ending. There is continuous demand of new strategies and deep insights for competitive edge to upkeep the company's image and its survival in the competitive market. This article is written keeping in mind how creativity and innovation are playing their role, and keeping their importance in the lives of managers.

Innovation is revealed by its Latin root nova or new. Innovation is the embodiment, combination or synthesis of knowledge in original, relevant, valued new products, processes or services. Thomas Edison's famous judgment that invention is 99% perspiration and 1% inspiration and this leads to innovation. Innovation involves acting on the creative ideas to make some specific and tangible difference in the domain in which the innovation occurs. Innovation can be explained as the successful implementation of creative ideas within an organization. In this view, creativity by individuals and teams is a starting point for innovation; the first is necessary but not sufficient condition for the second.

8.9.4   THE INNOVATION PROCESS

Any innovation process in any organization starts with an idea generation and how an opportunity is recognized out of the various ideas, but idea evaluation, development and finally the commercialization of the idea face a lot of obstacles, thus only few ideas make it to the end.

The Innovation Process

8.9.5   TYPES OF INNOVATION

8.9.5.1   INCREMENTAL INNOVATION

It is generally understood to exploit existing forms and technologies. It either improves upon something that already existed or reconfigures on existing form or technology to secure some other purpose.

8.9.5.2   RADICAL INNOVATION

A *radical innovation* is something new to the world, a departure from existing technology or methods. It is also known as breakthrough or discontinuous innovation, and so forth. The digital imaging technology used in today's consumers and professional cameras represents a radical departure from the chemically coated film technology (Fig. 8.1).

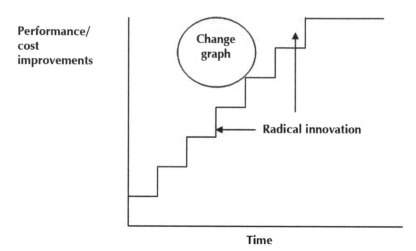

FIGURE 8.1   The process of innovation.

The course of innovation is generally characterized by long periods of incremental innovations penetrated by infrequent radical innovations.

For example, in electronics, we observe the introductions of vacuum tubes which were displaced by transistors, which were largely displaced by semi-conductors.

8.9.5.3   PRODUCT PROCESS INNOVATION

Creating an innovative new product at a price the target market will accept.

For example, in North America during the post war baby boom, when the market for baby diapers was huge they failed to gain market share. The reason was poor performance and price. When P&G entered R&D people quickly solved the performance problem using more suitable materials and a new design, although with a lot of difficulty, it was made cost effective, too.

8.9.5.4   SERVICE INNOVATIONS

This is another area in which innovation plays a key role. Great things happen when people rethink how best to serve customers, service innovation sometimes produces winning business models. For example, Dell's PCs are very good but—they share the same technologies as and offered by competitors what originally set this company apart and gave it a competitive edge was its innovative strategy of skipping the middlemen and selling customer—configured PCs directly to buyers. Later innovations in supply-claim management made this strategy fast and, effective and made Dell the world's most successful PC maker.

8.9.5.5 OPEN MARKET INNOVATION

Companies reach outside for the ideas they need for new products of services.

8.9.6 S-CURVE

Any organization can check whether it is innovative or not, meaning: moving with the trends in the market, is able to cope with the competition, and can stay in the market for long while by placing itself on the S-curve. Most companies resist new changes or innovations. A negative view of new ideas prevails that says that the new products do not serve the existing customers, but the company may not realize that by serving it to existing customers, there is a possibility that the new product will be just the ticket for new growth for the organization. The existing technology first settles and then becomes stagnant after reaching its peak and is overtaken by a new rival technology; thus, it becomes mandatory for every organization to encourage innovation. The S-curve is a measure of the speed of adoption of an innovation. It was first used by in 1903 by Gabriel Tarde, who first plotted the S-shaped diffusion curve.

There are four stages of the S-curve that have been reported, and it is important to note that innovation is different at each stage. The stages are:
1. Introduction
2. Growth
3. Maturity
4. Decline or Discontinuity

At the introductory stage, a great amount of effort, money, and resources are required, which in turn give mediocre or negligible performance improvements. At the growth stage, better know-how about the technology and processes are gathered, and progress is improved. By the time all the hindrances are taken care of, the innovation can be accepted and an increased growth will be observed,, and with relatively less effort and small resources, there may be high performance gains. At the end of the S-curve, when the technology starts to approach termination, the increase in performance becomes increasingly difficult. The S-curve is a powerful yet adaptable framework to understand and examine the various stages of innovations and to comprehend the technological cycles. The S-curve is used in assessing the performance in regard to time or effort. It also helps in understanding the current difficulties (which happen to be very common to products or services in certain phases of maturity) and how they can be avoided or minimized.

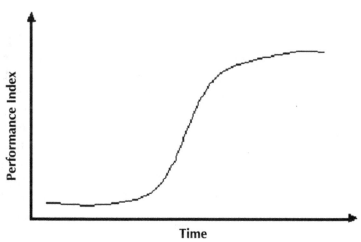

FIGURE 8.2   S curve

8.9.7 PROBLEMS

The decision makers normally find answers to the following questions when an opportunity is recognized:

1. Will the idea work?
2. Does the company have the technical know-how to make it work?
3. Does the idea represent value to customers?
4. Does the idea fit well with company's strategy?

5. Does it make sense from a cost perspective?
6. Radical Innovation may upset the industries or organizations existing business model. Disruptive technologies displace the established technologies.
7. The period of innovation is confirming and in certain to both producers of customers.
8. In the absence of technical standards, producers don't know which of several new courses to follow?
9. Customers are penalized over the choice of staying with the old technology, switching to the new, or waiting for dust to settle.
10. How might these innovations affect your own company's sales to profitability.
11. Radical Innovations are risky, expensive and time consuming.
12. Existing customers may preserve the company to stay in the old business.
13. The company's culture doesn't favor innovations.

The top management doesn't find the new innovation in the market to be a real threat. The new business is evaluated by managers of the mainstream business and by their standards which are profit of efficiency. The day-to-day concerns of those managers are diametrically opposed to those required by the innovative business, and they land up saying "this is drawing investment funds from our profitable businesses."

8.10 CREATIVITY

Creativity plays a critical role in the Innovation process. It sparks the initial idea and helps to improve the idea as it moves forward. An idea is recognized when a prepared mind is working on an experiment and "Chance favors the prepared mind,". This can be illustrated by the invention of various innovative products around the globe which were never an outcome of planned research or thoughtful process. 3M's fabric protector, and scotchgard, whose development was by virtue of an accident in 1953 in a laboratory. As available through resources and literature, one of its researcher was conducting a polymer experiments when a lab assistant accidentally spilled some of the solution on her shoes and she could not remove the spilled solution. As a result of her curiosity she realized that this ccould be something which may protect textiles from stains. Further development of the substance led to a successful new product. The stories are numerous and they have give the world new products not only useful to the society but also one of its own kind, Starting from the Thomson Edison invented electric bulb in 1892 to Google's innovations to date. Google is known for pushing technological boundaries, and it has certainly paid off. For innovation the following two quotes explains it all….

"There is a way to do it better—find it."
- Thomas Edison

"We should be focused on building things that don't exist."
- Larry Page, Google Co-Founder

8.10.1   BLOCKS TO CREATIVITY: BUILDING BLOCKS TO CREATIVITY

Resource myopia (near sightedness)	Resource fullness
Following the rules too closely too often	Ability to think outside the ruler
Fear of failure	Ability to accept failure of learn from it
Seeing play as only frivolous	Playfulness
Focusing on just the right answer	Focus on exploring possibilities
Being judgmental, critical	Being accepting
Discomfort with taking risk	Intelligent risk taking
Difficulty hearing another's prospective or opinion	Active listening, acceptance of differences
Lack of openers to ideas	Receptivity to ideas
Political problems and high better	Collaboration focus on mutual gains
Avoid ambiguity	Tolerance for ambiguity
Intolerance	Tolerance
Lack of flexibility	Flexibility
Going up too soon	Persistence
Worrying too much about what people will think	Having an inner focus
Thinking you're not creative	Recognizing creative potential in self

8.11   WAYS TO PROMOTE CREATIVITY AND INNOVATION IN ORGANIZATION

8.11.1   REWARDS

Those who generate ideas with pay or promotions or both provide a clear signal that good ideas are important. Monetary rewards appear to be more effective when they are performance based and when they give employee a personal stake in organizational successor rewards for innovators, however, should encompass more than pay of promotions. Monetary rewards of promotion prevent feelings of abuse—People aren't afraid to buy or suggest new things.

8.11.2   HIRE INNOVATIVE PEOPLE

Another effective way of promoting innovation is to hire right people with innovative bend of mind. Here are a few indicators to consider while hiring such people:

- Expert in one or two fields
- Enjoy doing immovable work
- Usually individual contributors
- Good problem solver
- Find new and different ways of seeing things

8.11.3   RECOGNITION

Acknowledging individual or group achievements in public forums that is, public announcement. This brings visible recognition for the contributor and motivates others to actively look for new and creative ideas.

8.11.4   CONTROL

Allowing an individual or group to participate in decision making or giving the individual or group, the resources needed to carry out a project. Access to greater resources is also an effective reward when innovation the goal.

8.12   CULTURE OF INNOVATION

A culture of innovation is encouraged by sending a clear message that the well-being of the company employees depends on continuous innovation.

Ideas of knowledge produce little when they are isolated in original pockets but something magical happens when those pockets are opened formally and isolated ideas come together to produce real opportunities. This can be done by cross pollution of ideas which can be promoted by HR.

- Periodically reasoning technical specialists
- Different work teams
- Send people to professional and scientific conferences
- Set up an intracompany knowledge management system—This makes knowledge and experience captured in one area available to everyone
- Sponsor events that bring outside experts to your company to give lectures of workshops—What they have to share often catalyzes ideas within the company
- Arrange periodic customer site visits
- Arrange field trips to observe best practices in other industries
- Meet with local inventors and entrepreneurs in your field
- Seek out consultants with different perspectives
- Seek out university professors on sabbatical to temporary join your group or participate in support sessions for innovators

8.13   ROLE OF HR PERSON

Typically, the market potential of innovations seems too small relative to the size of the existing business. Big companies miss out on many important innovations because the potential is often erroneously viewed as too small.

Human resource professional should provide freedom and support to the new innovative ideas by keeping the strategies and long-term objectives of the organization in mind at the same time giving freedom. Manager should be given specific directions about ends, but leave the means to employees.

There are two things that HR professionals can do to promote innovation: culture of innovation.

8.14   CULTURAL BARRIERS TO INNOVATION

8.14.1   HR AND CULTURE

The most powerful source for any organization is culture and in corporate culture, development is not a sole responsibility of the HR leadership but they certainly have a huge effect on creating the impact for culture inculcation amongst the employees. Culture also defines an organization and in organizations the culture is termed as organizational culture. According to Marguardt (2002), culture is an organization's values, beliefs, practices, rituals, and customs. Every organization has its own culture and as Wallace et al., 1999; define it to be "An organizational culture is understood as a characteristic of the day-to-day environment as seen and felt by those who work there."

Organizational culture is a binding force that holds the organization together and stimulates employees to commit to the organization and to perform. In present scenario, the organizational culture as defined by Deal and Kennedy (1982), Peters and Waterman (1982), and Schein and Edger (1990), culture is widely understood as an instrument to be used by management to shape and control in some way the beliefs, understandings and behaviors of individuals, and thus the organization to achieve specified goals.

In today's organizations it is believed that the strong organizational culture can be a source of organization's growth, whereas sometimes it can be a major resistance to incorporate change. In such situations, the role of HR management in managing the organizational culture in the best way HR plays an important role in developing a high performance organizational culture and thereby helps in achieving organizational objectives in an effective and efficient manner.

The employees learn culture through different forms such as:

- Stories: They contain a narrative of events about the organization's founders, key decisions that affect the organization's future course, and the present top management.
- Rituals: Activities such as recognition and award ceremonies
- Material symbols: The design and physical layout of spaces and buildings, furniture.
- Language

8.14.2 TRAIN AND REWARD FOR INNOVATION

Rewards have always been a positive motivator for reinforcing commitment, directing employee professional growth, and shaping the corporate culture to be more innovative. The HRD must work towards the implementation of effective reward mechanisms to develop the employees and culture of the organization. This can include compensation strategies, recognition, and reward programs.

8.14.3 THE INNOVATION LANDSCAPE MAP

An *Innovation Landscape* approach seeks to create an environment where innovation can be supported, interact, assemble and experiment anywhere and everywhere. The *Innovation Landscape Framework* is intended to promote better comprehension of the rich array of continuous and future efforts across an institution that supports innovation. The landscape helps in structuring the process which explores opportunities to start innovation and subsequently support approaches to obtain the desired impact. It is a tool, which can be used, and flourish further in various institutional and planning situations. It encourages the coordinated initiatives, which can be utilized effectively for an integrated framework for implementation.

The Innovation Landscape Map consists of four quadrants

1. The routine innovation
2. Disruptive innovation
3. Radical innovation
4. Architectural innovation

When organizations device an innovation strategy they have an option to decide about mobilizing their resources in business model innovations and how substantially it needs to focus on technological innovation. The Innovation Landscape Map considers how a possible and probable innovation unites with a company's existing business model and technical capabilities can assist with that decision.

8.14.3.1 HOW WILL INNOVATION CREATE VALUE FOR POTENTIAL CUSTOMERS?

Unless innovation induces potential customers to pay more, saves them money, or provides some larger societal benefit such as improved health or cleaner water, it is not creating value. Of course, innovation can create value in many ways. It might make a product perform better or make it easier or more convenient to use, more reliable, more durable, cheaper, and so on. Choosing what kind of value your innovation will create and then sticking to that is critical, because the capabilities required for each are quite different and take time to accumulate. For instance, Bell Labs created many diverse breakthrough innovations over a half century: the telephone exchange switcher, the photovoltaic cell, the transistor, satellite communications, the laser, mobile telephony, and the operating system Unix, to name just a few. But research at Bell Labs was guided by the strategy of improving and developing the capabilities and reliability of the phone network. The solid-state research program—which ultimately led to the invention of the transistor— was motivated by the need to lay the scientific foundation for developing newer, more reliable components for the communications system. Research on satellite communications was motivated in part by the limited bandwidth and the reliability risks of undersea cables. Apple consistently focuses its innovation efforts on making its products easier to use than competitors' and providing a seamless experience across its expanding family of devices and services. Hence its emphasis on integrated hardware–software development, proprietary operating systems, and design makes total sense.

8.15 CONCLUSION

Robinson and Stern (1997) contributed to the understanding of creativity in organizations by stating, in relation to creative acts: "Nobody can predict who will be involved in them, what they will be, when they will occur, or how they will occur" (p. 12).

However, they suggested six actions that are likely to increase the probability of creativity occurring.

1. Alignment to the organization's key goals

2. Opportunity for self-initiated activity—people are able to select a problem they are interested in solving
3. Opportunity/support for unofficial activity
4. Serendipity
5. Diverse stimuli
6. Good communication

According to Francisca Castro, Jorge Gomes, and Fernando C de Sousa

- Leader's emotional intelligence and employee creativity are positively associated.
- Work environment does not act as mediating factor between leadership and employee creativity.
- Instead leaders play a paramount role in stimulating employee creativity.

A study by Tierney, Farmer and Graen suggests that employees with a creative orientation: "work best under conditions that permit risk taking, operational autonomy, and the freedom to deviate from the status quo. Leader expression of enthusiasm or acceptance for innovation is one of the noted factors necessary for employees' motivation to be creative" (Tierney et al. 1999).

It is important to understand that creativity and innovation can emerge from anywhere: it may be from any of the stakeholders-employee customers groups, etc. The ranges of ideas vary and are unique in them. The ideas generated therefore bring in them the uniqueness, freshness and are important to survive in the world of competitiveness.

In the recent times the creativity and innovation has led to success and there are many success stories to be quoted and learn from. The creativity works not only in the marketing but also in the other areas like human resource management. The managers are constantly work to improve on the policies to make them innovative and creative.

Stimulating creativity and exploring new ideas result in encouraging the employees to think out of the box and giving them an edge over the competition. Creativity improves the problem solving process. It gives a competitive edge over any business which is striving to be successful.

It is important to have a free exchange of ideas to be supported and encouraged by the company.

KEYWORDS

- creativity
- innovation
- S curve
- HR practices

REFERENCES

Amabile, T. M. A model of Creativity and Innovation in Organizations. In *Research in Organizational Behavior*; Staw, B. M., Cummings, L. L., Eds.; Press: Greenwich, CT, 1988; Vol. 10, pp. 123–167.

Amabile, T. M. *Creativity in Context*. Westview Press: Boulder, CO, 1996.

Amabile, T. M. Motivating Creativity in Organizations: On Doing What You Love and Loving What You Do. *Calif. Manage. Rev.* **1997**, *40*(1)39–58.

Amabile, T., & Khaire, M. (2008). *Creativity and the role of the leader*. [Boston, MA]: [Harvard Business School Publishing].

Constantine A. (2001) Determinants of Organisational Creativity: A Literature Review. *Manage. Decis.* **2001**, *39*(10)834–841.

Giustiniano, L.; Lombardi, S.; Cavaliere, V. Knowledge Collecting Fosters Organizational Creativity. *Manage. Decision* **2016**, *54*(6)1464–1496.

Hamel, G., Prahalad, C., & Välikangas, L. (2006). *Staying ahead of your competition*. [Boston, MA]: Harvard Business School Publishing].

Luecke, R., & Watkins, M. (2003). Harvard business essentials. Boston, Mass.: Harvard Business School Press.

Marquardt, M. J. (2002). Building the learning organization: Mastering the 5 elements for corporate learning (2nd ed.). Palo Alto, CA: Davies-Black Publishing, Inc.

Mumford, M, D.; Gustafson, S. B. Creativity Syndrome: Integration, Application, And Innovation. *Psychol. Bull.* **1988**, *103*, 27–43.

Paulus, P. B.; Nijstad, B. A (2003). Group Creativity: Innovation through Collaboration Oxford University Press.

Woodman, R, W.; Sawyer, J. E.; Griffin, Ricky W. *Acad. Manage. Rev.* **1993**, *18*(2)293–321.

Wallace, J.; Hunt, J.; Richards, C. The Relationship between Organisational Culture, Organisational Climate and Managerial Values. *Int. J. Public Sector Manage.* **1999**, *12*, 548–564.

http://innovationexcellence.com/blog/2012/08/04/whats-the-difference-between-creativity-and-innovation/.

https://hbr.org/2005/07/managing-for-creativity.

INDEX